AF360706

Advances in Organic Coatings 2018

Advances in Organic Coatings 2018

Editor

Flavio Deflorian

MDPI • Basel • Beijing • Wuhan • Barcelona • Belgrade • Manchester • Tokyo • Cluj • Tianjin

Editor
Flavio Deflorian
Department of Industrial Engineering,
University of Trento
Italy

Editorial Office
MDPI
St. Alban-Anlage 66
4052 Basel, Switzerland

This is a reprint of articles from the Special Issue published online in the open access journal *Coatings* (ISSN 2079-6412) (available at: https://www.mdpi.com/journal/coatings/special_issues/adv_org_coat).

For citation purposes, cite each article independently as indicated on the article page online and as indicated below:

LastName, A.A.; LastName, B.B.; LastName, C.C. Article Title. *Journal Name* **Year**, *Article Number*, Page Range.

ISBN 978-3-03936-607-1 (Hbk)
ISBN 978-3-03936-608-8 (PDF)

Contents

About the Editor

Flavio Deflorian is a Full Professor in Materials Science and Technology at the Department of Industrial Engineering of the University of Trento. After his Master's in Materials Engineering at the University of Trento Italy and Master's in Advanced Materials Technology, University of Surrey, Guidford U.K., he received his PhD in Materials Engineering from the University of Bologna, Italy. His scientific activity is reported in more than 180 papers published in international journals, and in about 20 books chapters and 2 books. He is also the co-author of more than 150 conference presentations. His main research fields are related to surface engineering for material durability and electrochemical techniques for material characterization. His h-index is 40.

Preface to "Advances in Organic Coatings 2018"

The scientific and technological advances in organic coatings have been impressive over the last couple of decades, and recent further developments are opening new prospective for organic coatings science and technology. New materials, based on nanotechnologies (nanostructured polymeric matrixes, nano-pigments), and new surface pretreatments improving the chemical and physical stability of the interfaces, are deeply modifying the performances of organic coatings. Moreover, organic coatings are increasingly multifunctional. In addition to traditional functions, such as corrosion protective actions or aesthetical functions, modern organic coatings must often support additional roles: antibacterial activity, self-healing ability, and tribological properties, etc.

The recent advances in experimental techniques (electrochemical methods, optical and electron microscopy, chemical surface analysis, and thermal analysis, etc.) applied to organic coatings provide a powerful tool for research and scientific development in this area. Further driving forces pulling innovation into the organic coatings area are environmental issues. In order to develop new systems, joining advanced performance with high environmental sustainability, new materials are under development, anticipating future legislative requirements. This frame is induced to consider the advances in organic coatings (the skin of materials) as one of the most interesting and promising innovation fields in material science.

The aim of this Special Issue is to provide an update on the most advanced research in this area, showing the innovation trends and promoting further research for better properties for new coatings materials. The eight papers composing the Special Issue offer an interesting and broad overview of the developments in the science and technology of organic coatings from various and differentiated points of view.

Some works focus on a key point in the case of organic coatings: the interface—both the interface between the coating and the external environment (the surface of the material), and the interface between the coating and the substrate. The paper of Wasser et al. describes the development of bio-inspired fluorine free synthetic replica of lotus and nasturtium surfaces in order to obtain self-cleaning properties. Hydrophobicity is also the topic of the paper proposed by Jankauskaitė et al., where the improvement of the non-wetting properties of a flexible polydimethylsiloxane substrate surface via a plasma-polymerized hexamethyldisilazane thin film deposition by the arc discharge method is described. Considering the substrate-coating interface, Nabavian et al. present a paper assessing the influence of benzoimidazole concentrations on the cathodic delamination of epoxy coating applied on steel substrate. The Special Issue also presents interesting experimental work concerning newly formulated coatings for special applications: edible coatings, flame-retardant coatings, and cultural heritage protection coatings.

Kozakiewicz et al. present a simultaneous synthesis of aqueous silicone-acrylic and acrylic-silicone hybrid dispersions by different emulsion polymerizations. The hybrid dispersions showed good mechanical stability, narrow particle size distribution, and tended to form mechanically strong continuous coatings and films. Bernardino-Nicanor et al. studied Opuntia Robusta Mucilage as a possible edible coating. FTIR analysis demonstrated important differences in the mucilage extracted from different tissues in terms of pectic content. The paper studying a new stratified flame-retardant coating, is presented by Beaugendre et al.. A polymer blend composed of silicone of a curable epoxy resin and of a liquid functional filler successfully formed a double-layered coating, showing excellent adhesion to the polycarbonate substrate.

Roncagliolo Barrera et al. presented a paper characterizing organic coatings for cultural heritage artefact conservation using electrochemical noise. This technique is capable of assessing with great sensitivity the increase in corrosion-resistance performance conferred by the inhibitors added to protective coating. The addition of nicotine and caffeine to temporary protection demonstrated a high protection efficacy (that was better for nicotine than caffeine in acrylic coatings).

Finally, an important topic related to circular economy and material recycling is presented in the paper by Silva et al., where the properties of post-consumer polyethylene terephthalate (PET) coating mechanically deposited on mild steels were presented.

This Special Issue is a helpful tool for researchers working in the field, for appreciating the state of the art, and to provide an overview of the new innovative trends in the field of organic coatings science and technology.

Flavio Deflorian
Editor

Editorial

Special Issue: "Advances in Organic Coatings 2018"

Flavio Deflorian

Department of Industrial Engineering, University of Trento, via Sommarive n. 9, 38123 Trento, Italy; flavio.deflorian@unitn.it

Received: 3 June 2020; Accepted: 9 June 2020; Published: 10 June 2020

Abstract: Organic coatings have shown an impressive evolution in recent years, both scientifically and technologically. Nanotechnology and surface science allows the development of multifunctional materials combining different properties, such as corrosion protective actions, aesthetical functions, hydrophobic properties, and self-healing ability. In addition, recent advances in experimental techniques and the attention to environmental issues are pushing to develop new systems, joining advanced performance with high sustainability. The aim of this Special Issue is to provide an update on the most advanced research in this area, showing the innovation trends and promoting further research for better properties of new coatings materials.

Keywords: flame-retardant coatings; corrosion protection; hydrophobic coatings; self-cleaning; bio-inspired materials

---

The scientific and technological advances in organic coatings have been impressive over the last couple of decades, and recent further developments are opening new prospective for organic coatings science and technology [1]. New materials, based on nanotechnologies (nanostructured polymeric matrixes, nano-pigments), and new surface pretreatments improving the chemical and physical stability of the interfaces, are deeply modifying the performances of organic coatings [2–4]. Moreover, organic coatings are increasingly multifunctional. In addition to traditional functions, such as corrosion protective actions or aesthetical functions, modern organic coatings must often support additional roles: antibacterial activity, self-healing ability, and tribological properties, etc. [5]

The recent advances in experimental techniques (electrochemical methods, optical and electron microscopy, chemical surface analysis, and thermal analysis, etc.) applied to organic coatings provide a powerful tool for research and scientific development in this area [6]. Further driving forces pulling innovation into the organic coatings area are environmental issues. In order to develop new systems, joining advanced performance with high environmental sustainability, new materials are under development, anticipating future legislative requirements [7]. This frame is induced to consider the advances in organic coatings (the skin of materials) as one of the most interesting and promising innovation fields in material science.

The aim of this Special Issue is to provide an update on the most advanced research in this area, showing the innovation trends and promoting further research for better properties for new coatings materials. The eight papers composing the Special Issue [8–15] offer an interesting and broad overview of the developments in the science and technology of organic coatings from various and differentiated points of view.

Some works focus on a key point in the case of organic coatings: the interface—both the interface between the coating and the external environment (the surface of the material) [8,15], and the interface between the coating and the substrate [12]. The paper of Wasser et al. [15] describes the development of bio-inspired fluorine free synthetic replica of lotus and nasturtium surfaces in order to obtain self-cleaning properties. A UV curable acrylate oligomer and acrylated siloxane comonomers were used, combining the self-assembly of the comonomers and a replication step of the plant surfaces.

The texturization of the surfaces had a large influence on hydrophobicity and, as a result, a self-cleaning surface was obtained with very attractive application possibilities.

Hydrophobicity is also the topic of the paper proposed by Jankauskaitė et al. [8], where the improvement of the non-wetting properties of a flexible polydimethylsiloxane substrate surface via a plasma-polymerized hexamethyldisilazane thin film deposition by the arc discharge method is described. The film is composed of nanoparticles forming a branched network with self-cleaning and non-wetting behavior, demonstrating a new strategy for the large-scale fabrication of superhydrophobic surfaces with a self-cleaning function on flexible substrates.

Considering the substrate-coating interface, Nabavian et al. present a paper [12] assessing the influence of benzoimidazole concentrations on the cathodic delamination of epoxy coating applied on steel substrate. The results demonstrated that the cathodic disbonding resistance of epoxy-polyamide coating was dependent on the inhibitor content. Moreover, the interface stability, measured as the wet adhesion of the polymeric coating, was significantly enhanced through the addition of 0.75 wt.% benzoimidazole.

The Special Issue also presents interesting experimental work concerning newly formulated coatings [10] for special applications: edible coatings [13], flame-retardant coatings [14], and cultural heritage protection coatings [11].

Kozakiewicz et al. [10] present a simultaneous synthesis of aqueous silicone-acrylic and acrylic-silicone hybrid dispersions by different emulsion polymerizations. The hybrid dispersions showed good mechanical stability, narrow particle size distribution, and tended to form mechanically strong continuous coatings and films. The selected hybrid dispersions described in this paper can be applied as binders in the formulation of architectural paints that will be characterized by high water resistance and high surface hydrophobicity combined with high water vapor permeability.

Bernardino-Nicanor et al. [13] studied Opuntia Robusta Mucilage as a possible edible coating. FTIR analysis demonstrated important differences in the mucilage extracted from different tissues in terms of pectic content. In conclusion, this study showed that Opuntia Robusta Mucilage is a promising edible coating and that the tissue and extraction solvent influences mucilage characteristics.

The paper studying a new stratified flame-retardant coating, is presented by Beaugendre et al. [14]. A polymer blend composed of silicone of a curable epoxy resin and of a liquid functional filler successfully formed a double-layered coating, showing excellent adhesion to the polycarbonate substrate. The two phosphorus-based compound fillers were added to improve the flame-retardant properties.

Roncagliolo Barrera et al. [11] presented a paper characterizing organic coatings for cultural heritage artefact conservation using electrochemical noise. This technique is capable of assessing with great sensitivity the increase in corrosion-resistance performance conferred by the inhibitors added to protective coating. The addition of nicotine and caffeine to temporary protection demonstrated a high protection efficacy (that was better for nicotine than caffeine in acrylic coatings).

Finally, an important topic related to circular economy and material recycling is presented in the paper by Silva et al. [9], where the properties of post-consumer polyethylene terephthalate (PET) coating mechanically deposited on mild steels were presented. The mechanical deposition (press recycled) of the coating does no significant damage to the polymer and the PET organic coating presents good adhesion to the substrate as well as a high corrosion protection for carbon steel during long immersion times in aggressive solution. The PET organic coatings can be considered as an alternative, both for the PET recycling and for a new anticorrosive coating.

This Special Issue is a helpful tool for researchers working in the field, for appreciating the state of the art, and to provide an overview of the new innovative trends in the field of organic coatings science and technology.

Conflicts of Interest: The author declares no conflict of interest.

References

1. Jones, F.N.; Nichols, M.E.; Pappas, S.P. *Organic Coatings: Science and Technology*; John Wiley & Sons Inc.: Hoboken, NJ, USA, 2017.
2. Figueira, R.B.; Silva, C.J.R.; Pereira, E.V. Organic-inorganic hybrid sol-gel coatings for metal corrosion protection: A review of recent progress. *J. Coat. Technol. Res.* **2015**, *12*, 1–35. [CrossRef]
3. Si, Y.; Guo, Z. Superhydrophobic nanocoatings: From materials to fabrications and to applications. *Nanoscale* **2015**, *7*, 5922–5946. [CrossRef] [PubMed]
4. Hornberger, H.; Virtanen, S.; Boccaccini, A.R. Biomedical coatings on magnesium alloys—A review. *Acta Biomater.* **2012**, *8*, 2442–2455. [CrossRef] [PubMed]
5. Montemor, M.F. Functional and smart coatings for corrosion protection: A review of recent advances. *Surf. Coat. Technol.* **2014**, *258*, 17–37. [CrossRef]
6. Huang, V.M.; Wu, S.; Orazem, M.E.; Pebere, N.; Tribollet, B.; Vivier, V. Local electrochemical impedance spectroscopy: A review and some recent developments. *Electrochim. Acta* **2011**, *56*, 8048–8057. [CrossRef]
7. Ataei, S.; Khorasani, S.N.; Neisiany, R.E. Biofriendly vegetable oil healing agents used for developing self-healing coatings: A review. *Prog. Org. Coat.* **2019**, *129*, 77–95. [CrossRef]
8. Jankauskaitė, V.; Narmontas, P.; Lazauskas, A. Control of Polydimethylsiloxane Surface Hydrophobicity by Plasma Polymerized Hexamethyldisilazane Deposition. *Coatings* **2019**, *9*, 36. [CrossRef]
9. Silva, E.; Fedel, M.; Deflorian, F.; Cotting, F.; Lins, V. Properties of Post-Consumer Polyethylene Terephthalate Coating Mechanically Deposited on Mild Steels. *Coatings* **2019**, *9*, 28. [CrossRef]
10. Kozakiewicz, J.; Trzaskowska, J.; Domanowski, W.; Kieplin, A.; Ofat-Kawalec, I.; Przybylski, J.; Woźniak, M.; Witwicki, D.; Sylwestrzak, K. Studies on Synthesis and Characterization of Aqueous Hybrid Silicone-Acrylic and Acrylic-Silicone Dispersions and Coatings. Part I. *Coatings* **2019**, *9*, 25. [CrossRef]
11. Roncagliolo Barrera, P.; Rodríguez Gómez, F.J.; García Ochoa, E. Assessing of New Coatings for Iron Artifacts Conservation by Recurrence Plots Analysis. *Coatings* **2019**, *9*, 12. [CrossRef]
12. Nabavian, S.; Naderi, R.; Asadi, N. Determination of Optimum Concentration of Benzimidazole Improving the Cathodic Disbonding Resistance of Epoxy Coating. *Coatings* **2018**, *8*, 471. [CrossRef]
13. Bernardino-Nicanor, A.; Montañez-Soto, J.L.; Conde-Barajas, E.; Xochilt Negrete-Rodríguez, M.L.; Teniente-Martínez, G.; Vargas-León, E.A.; Juárez-Goiz, J.M.S.; Acosta-García, G.; González-Cruz, L. Spectroscopic and Structural Analyses of Opuntia Robusta Mucilage and Its Potential as an Edible Coating. *Coatings* **2018**, *8*, 466. [CrossRef]
14. Beaugendre, A.; Degoutin, S.; Bellayer, S.; Pierlot, C.; Duquesne, S.; Casetta, M.; Jimenez, M. Self-Stratification of Ternary Systems Including a Flame Retardant Liquid Additive. *Coatings* **2018**, *8*, 448. [CrossRef]
15. Wasser, L.; Dalle Vacche, S.; Karasu, F.; Müller, L.; Castellino, M.; Vitale, A.; Bongiovanni, R.; Leterrier, Y. Bio-Inspired Fluorine-Free Self-Cleaning Polymer Coatings. *Coatings* **2018**, *8*, 436. [CrossRef]

Communication

Control of Polydimethylsiloxane Surface Hydrophobicity by Plasma Polymerized Hexamethyldisilazane Deposition

Virginija Jankauskaitė [1],*, Pranas Narmontas [2] and Algirdas Lazauskas [2]

[1] Department of Production Engineering, Kaunas University of Technology, Studentu St. 56, LT-51424 Kaunas, Lithuania

[2] Institute of Materials Science, Kaunas University of Technology, Barsausko St. 59, LT-51423 Kaunas, Lithuania; pranas.narmontas@ktu.lt (P.N.); algirdas.lazauskas@ktu.lt (A.L.)

* Correspondence: virginija.jankauskaite@ktu.lt; Tel.: +370-610-05-190

Received: 30 November 2018; Accepted: 10 January 2019; Published: 11 January 2019

Abstract: The properties of a polydimethylsiloxane (PDMS) surface were modified by a one-step deposition of plasma polymerized hexamethyldisilazane (pp-HMDS) by the arc discharge method. Scanning electron microscopy, atomic force microscopy, and Fourier-transform infrared spectroscopy analytical techniques were employed for morphological, structural, and chemical characterization of the pp-HMDS modified PDMS surface. The changes in PDMS substrate wetting properties were evaluated by means of contact angle measurements. The unmodified PDMS surface is hydrophobic with a contact angle of 122°, while, after pp-HMDS film deposition, a dual-scale roughness PDMS surface with contact angle values as high as 170° was obtained. It was found that the value of the contact angle depends on the plasma processing time. Chemically, the pp-HMDS presents methyl moieties, rendering it hydrophobic and making it an attractive material for creating a superhydrophobic surface, and eliminating the need for complex chemical routes. The presented approach may open up new avenues in design and fabrication of superhydrophobic and flexible organosilicon materials with a self-cleaning function.

Keywords: polydimethylsiloxane; superhydrophobicity; hexamethyldisilazane; plasma polymerization

---

1. Introduction

Polydimethylsiloxanes (PMDS) are the most widely used silicon-based organic polymers, commonly referred to as silicones. Because of easy fabrication, non-toxicity, biocompatibility and biodurability they have found potential applications in various fields. The surface of PDMS is naturally hydrophobic, but a number of efforts have been made to modify PDMS and further enhance its hydrophobicity [1,2]. PDMS hydrophobicity plays an important role in diverse applications e.g., self-cleaning surfaces [3], microfluidics [4], microelectromechanical systems [5], and biomedical applications [6].

Superhydrophobic PDMS surfaces can be fabricated by pulsed laser irradiation resulting in surface modification with a static contact angle (CA) value of 170° [7]. However, the whole irradiation procedure is highly time-consuming, thus limiting the scalability of this method. A more reliable and effective practice includes the deposition/formation of a thin film on the surface of the material to obtain the desired functionality. Plasma treatment is attractive as the processing time is short, the process involves low temperature, and procedures are relatively simple. Importantly, a single-step technique is desired for obtaining superhydrophobic and self-cleaning surface functionalities.

In this contribution, we fabricated a superhydrophobic PDMS surface via plasma polymerized hexamethyldisilazane (pp-HMDS) thin film deposition by arc discharge. To the best of our knowledge, the technique adopted here has been not reported for the fabrication of superhydrophobic PDMS surface using hexamethyldisilazane monomer as a precursor. HMDS is well known as being widely used for hydrophobic coatings on various hydroxyl-bearing surfaces [8,9]. HMDS chemical activity derives from the presence of a highly reactive nitrogen atom within the compound. The presented one step deposition of in situ polymerized hexamethyldisilazane is simple and scalable, and thus can provide a new strategy for the large scale fabrication of superhydrophobic surfaces with a self-cleaning function on flexible substrates.

2. Materials and Methods

The addition-curing silicone rubber Elastosil RT 601 A/B with a viscosity of 3500 mPa·s at 23 °C (Wacker Chemie AG, Múnich, Germany) was used as received for flexible films fabrication. HMDS of analytical grade ($\geq$99%, Sigma-Aldrich, Saint Louis, MO, USA) was used as received.

The experimental setup of arc plasma reactor and technological conditions have been reported previously [10]. Briefly, a rod-shaped graphite anode and cathode were placed at the center of the discharge chamber. A quartz cuvette containing HMDS solution was positioned 20 mm from the anode, and PDMS substrate was placed at a distance of 15 mm from the electrodes. The chamber was connected to a vacuum line backed by a rotary pump. Arc-discharge was generated between anode and cathode using a DC transferred arc process using ~4.3 mA current and ~25 kV voltage. The deposition time was varied up to 60 s.

A FEI Quanta 200 scanning electron microscope (SEM, Thermo Fisher Scientific, Waltham, MA, USA) was used to collect micrographs of the investigated surface. The samples were imaged at an accelerating voltage of 30 kV. Atomic force microscopy (AFM) experiments were carried out with NT-206 (Microtestmachines, New Taipei City, Taiwan) in air at room temperature (22 $\pm$ 1 °C) using a V-shaped silicon cantilever operating in contact mode. The surface morphology of the resulting films was evaluated based on the AFM surface topography images and roughness parameters. Vertex 70 Fourier transform infrared (FTIR) spectrometer (Bruker Optics Inc., Billerica, MA, USA) equipped with a 30Spec (Pike Technologies, Madison, IA, USA) specular reflectance accessory having a fixed 30° angle of incidence was used for the chemical characterization of the modified PMDS surface.

CA measurements were performed at room temperature using the sessile drop method. A droplet of deionized water (5 μL) was deposited onto the investigated surface. Optical images of the droplet were recorded with a PC-connected digital camera after 10 s of dropping and CA measurements were carried out using an active contour method based on B-spline snakes (active contours) [11]. The contact angle hysteresis was measured as the difference between the advancing and receding contact angle of a sliding droplet. The test was performed by setting a droplet on a sample, which was placed on a horizontal plate. The plate was tilted slowly until the water drop began to slide along the surface; at this point the camera shutter was activated. The advancing and receding contact angles where then measured.

3. Results

The morphology of unmodified and plasma polymerized HMDS (pp-HMDS) modified flexible PDMS substrate and water droplets on the PDMS surface before and after pp-HMDS film deposition at different times are compared in Figure 1. Cured PDMS is produced spontaneously forming wavy structures on the surface with micro-scale amplitude and periodicity of 128 nm. SEM images of the resulting pp-HDMS thin film surface for deposition times of 30 and 60 s are presented in Figure 1b,c, respectively. The deposition resulted in a highly branched and crosslinked pp-HMDS structures composed of quasi-spherical nanoparticles with size in the range of 15–60 nm. Growth in three-dimensional assembles and formation of large nanoparticles aggregates were observed as the deposition time increased.

These morphological alterations change the wetting properties of the PDMS surface (Figure 1). The unmodified PDMS surface exhibits hydrophobic behavior with a static CA value not higher than 122°.

A considerable improvement in non-wetting characteristics of pp-HMDS film functionalized surfaces was observed. After 30 s of deposition, the nanostructured pp-HMDS film exhibited superhydrophobic properties with static CA values of 169°–170°. In this case the low value of CA hysteresis (2°), defined as the difference between the CA at the front of the droplet (advancing CA) and at the back of the droplet (receding CA), was obtained. The increase of deposition times up to 60 s results in lower CA values, i.e., CA = 159°–161°.

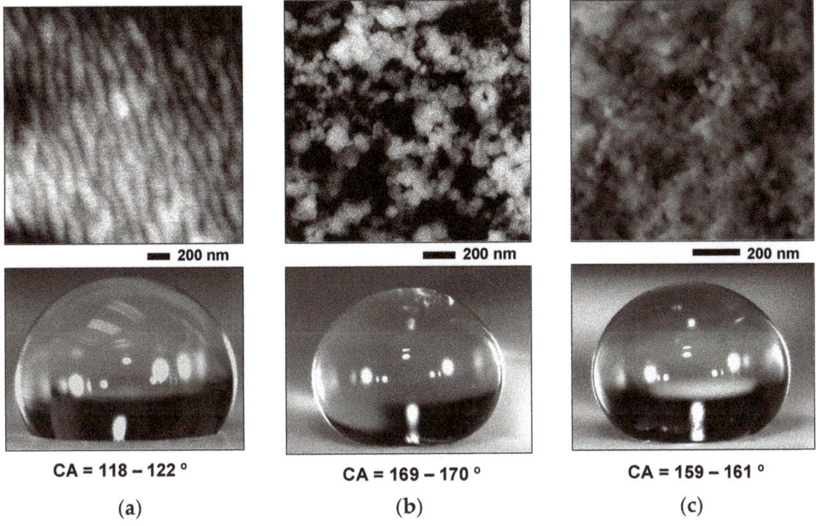

Figure 1. SEM image of unmodified (**a**) and pp-HMDS film modified PDMS surfaces (**b,c**) and water droplets on PDMS surface before and after pp-HMDS film deposition at different times: (**a**) 0; (**b**) 30 s; (**c**) 60 s.

The presence of nanostructures in the form of quasi-spherical nanoparticles and interconnection caused by the formation of large aggregates at longer pp-HMDS deposition time (60 s) was confirmed using characteristic AFM topographical and surface profile images shown in Figure 2. In this case dual-scale roughness of the surface was maintained, the pp-HMDS film surface was found to be rough with the root-mean square roughness having a value of 96.11 nm. However, the spiky surface morphology changes into a bumpy one and the negative surface skewness parameter value (-0.2) indicates predominance of valleys.

As can be seen from Figure 3a, in the FTIR absorbance spectrum of unmodified PDMS the bands at 2965 and 2906 cm^{-1} are assigned to asymmetric and symmetric stretching of CH$_3$ groups, respectively [12]. The asymmetric and symmetric bending vibrations of CH$_3$ groups are also observed at 1410 and 1258 cm^{-1}, respectively. The bands at 1072 and 1007 cm^{-1} are characteristic of Si–O–Si asymmetric and symmetric stretching vibrations, respectively. Asymmetric rocking at 864 cm^{-1} and stretching at 785 cm^{-1} vibrations can be attributed to the Si–CH$_3$ group [12].

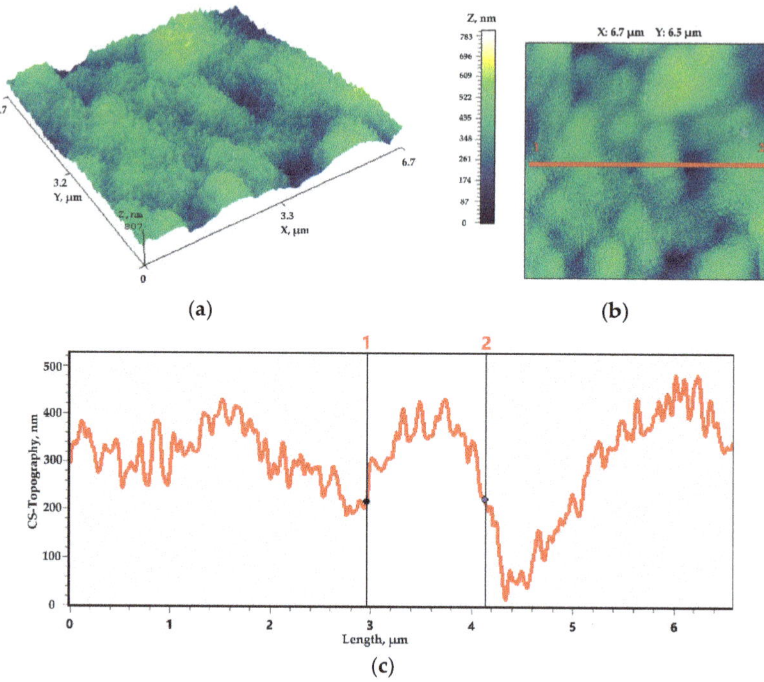

(a)   (b)

(c)

Figure 2. AFM image of 3D (**a**) and 2D (**b**) topography with normalized Z (nm scale), and profilogram of pp-HMDS film at deposition time of 60 s (**c**).

The deposition of polymerized pp-HMDS film leads to the obvious PDMS surface functional group changes (Figure 3b). The broad band between 3400 and 3700 cm^{-1} is related to O–H stretching in Si–OH bonds of hydrophilic silanol groups [13]. The absorbance at 3350 cm^{-1} is characteristic for stretching of the N–H bond, while the doublet at 2350 cm^{-1} is attributed to CO_2 species [10]. As in the case of unmodified PDMS, the presence of methyl moieties in the modified surface is confirmed by an absorption band at 1410 cm^{-1}, related to CH_3 asymmetric bending in Si–CH_3 bonds, and 2965 and 2906 cm^{-1} bands, which are characteristic for asymmetric stretching and symmetric stretching of the CH_3 group, respectively [14,15]. A low intensity band at 1454 cm^{-1} is assigned to the asymmetric bending vibrations of the CH_2 group in the Si–CH_2–CH_2–Si link that play a substantial role in the cross-linking process during HMDS polymerization [15]. The band located at 2250 cm^{-1} corresponds to Si–H stretching vibration [16], while the band at 1629 cm^{-1} can be assigned to stretching of C=O [17]. Some oxygen related functional groups could arise from free radical (possibly trapped in the film structure) reaction with the atmosphere, when the samples are removed from the reactor [18].

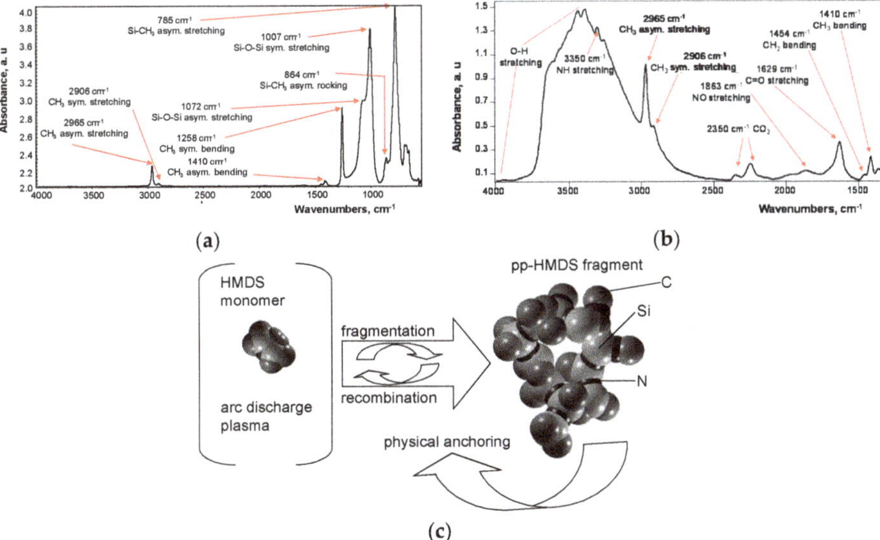

(a)   (b)

(c)

Figure 3. FTIR absorbance spectra of flexible PDMS substrate (**a**) and pp-HMDS (**b**) with functional groups assigned and schematic diagram of probable deposition mechanism (**c**); HMDS monomer and pp-HMDS network fragment are shown in van der Waal's-based representation.

4. Discussions

Generally, the hydrophobic properties of films are determined by the kind and amount of grafted hydrophobic groups and surface roughness parameters. One step coating by in situ HMDS deposition and polymerization is an easy and rapid method to impart non-wetting properties to the PDMS surface. The plasma polymerization process of HMDS monomers resulted in highly branched and crosslinked structures composed of quasi-spherical nanoparticles. After pp-HMDS film deposition, the PDMS surface shows superhydrophobic characteristics with a CA value close to 170°. High hydrophobicity of pp-HMDS originates from the high amount of CH_3 species and specific film surface morphology. The pp-HMDS film functionalized surfaces exhibited Cassie–Baxter state with a "lotus effect" observable and a low CA hysteresis of 2°, suggesting that a water droplet is not able to wet the spaces between surface morphological features allowing air pockets to remain at the interface. The increase of pp-HMDS film deposition time influences the decrease of CA value. It can be attributed to a higher solid fraction of surface morphological features in contact with the water droplet, which decreases the concentration of air pockets trapped at the interface of pp-HMDS with the droplet.

Based on the SEM, FTIR data, and surface wetting studies, it is suggested that the HDMS monomer molecules passing to the arc plasma region during the operational process are fragmented with partial retention and formation of new chemical bonds. The corresponding repetition of fragmentation and recombination reactions of HMDS monomers in arc plasma leads to the deposition of a randomly crosslinked network structure of pp-HMDS (Figure 3c) and forms a heterogeneous surface with a high fraction of methyl moieties retained, thus providing superhydrophobic characteristics with a self-cleaning function.

Thus, FTIR investigations reveal multiple non-covalent interaction achieved by in situ HMDS polymerization with physical anchoring on the polymer surface [19]. Such an interaction can be recognized as the driving force for constructing and fabrication of superhydrophobic and flexible organosilicon materials with a self-cleaning function.

5. Conclusions

Herein, we successfully enhanced the non-wetting properties of a flexible polydimethylsiloxane substrate surface via plasma polymerized hexamethyldisilazane thin film deposition by the arc discharge method. Such a film is composed of quasi-spherical nanoparticles stacked together, which form a branched network. The deposited nanostructured plasma polymerized hexamethyldisilazane film exhibits superhydrophobic properties with static contact angle values as high as 170° and a low contact angle hysteresis of 2°. The PDMS surface undergoes self-cleaning and non-wetting behavior due to the multiple non-covalent interactions attended by the incorporation in the surface layer of methyl groups and a nano-rough surface formation. This is a facile and effective method that can provide a new strategy for the large scale fabrication of superhydrophobic surfaces with a self-cleaning function on flexible substrates.

Author Contributions: Conceptualization, A.L. and V.J.; Methodology, A.L. and P.N.; Validation, V.J. and P.N.; Formal Analysis, V.J.; Investigation, A.L. and P.N.; Resources, A.L. and V.J.; Writing—Original Draft Preparation, V.J. and A.L.; Writing—Review & Editing, V.J. and A.L.; Visualization, A.L. and V.J.; Supervision, V.J.; Project Administration, A.L.; Funding Acquisition, V.J.

Funding: This research was funded by the European Social Fund under "Development of Competences of Scientists, other Researchers and Students through Practical Research Activities" measure (No. 09.3.3-LMT-K-712-01-0074).

Conflicts of Interest: The authors declare no conflict of interest.

References

1. Yang, J.; Pi, P.; Wen, X.; Zheng, D.; Xu, M.; Cheng, J.; Yang, Z. A novel method to fabricate superhydrophobic surfaces based on well-defined mulberry-like particles and self-assembly of polydimethylsiloxane. *Appl. Surf. Sci.* **2009**, *255*, 3507–3512. [CrossRef]
2. Slaughter, G.; Stevens, B. A cost-effective two-step method for enhancing the hydrophilicity of PDMS surfaces. *BioChip J.* **2014**, *14*, 28–34. [CrossRef]
3. Tavares, M.T.S.; Santos, A.S.F.; Santos, I.M.G.; Silva, M.R.S.; Bomio, M.R.D.; Longo, E.; Paskocimas, C.A.; Motta, F.V. TiO$_2$/PDMS nanocomposites for use on self-cleaning surfaces. *Surf. Coat. Technol.* **2014**, *239*, 16–19. [CrossRef]
4. Tropmann, A.; Tanguy, L.; Koltay, P.; Zengerle, R.; Riegger, L. Completely superhydrophobic PDMS surfaces for microfluidics. *Langmuir* **2012**, *28*, 8292–8295. [CrossRef] [PubMed]
5. Basu, B.J.; Bharathidasan, T.; Anandan, C. Superhydrophobic oleophobic PDMS-silica nanocomposite coating. *Surf. Innov.* **2013**, *1*, 40–51. [CrossRef]
6. Martin, S.; Bhushan, B. Transparent, wear-resistant, superhydrophobic and superoleophobic poly(dimethylsiloxane) (PDMS) surfaces. *J. Colloid Interface Sci.* **2017**, *488*, 118–126. [CrossRef] [PubMed]
7. Khorasani, M.T.; Mirzadeh, H. In vitro blood compatibility of modified PDMS surfaces as superhydrophobic and superhydrophilic materials. *J. Appl. Polym. Sci.* **2004**, *91*, 2042–2047. [CrossRef]
8. Terpilowski, K.; Goncharuk, O. Hydrophobic properties of hexamethyldisilazane modified nanostructured silica films on glass: Effect of plasma pre-treatment of glass and polycondensation features. *Mater. Res. Express* **2018**, *5*, 016409. [CrossRef]
9. Protsak, I.; Pakhlov, E.; Tertykh, V.; Le, Z.-C.; Dong, W. A new route for preparation of hydrophobic silica nanoparticles using a mixture of poly(dimethylsiloxane) and diethyl carbonate. *Polymers* **2018**, *10*, 116. [CrossRef]
10. Lazauskas, A.; Baltrusaitis, J.; Grigaliūnas, V.; Jucius, D.; Guobienė, A.; Prosyčevas, I.; Narmontas, P. Characterization of plasma polymerized hexamethyldisiloxane films prepared by arc discharge. *Plasma Chem. Plasma Process.* **2014**, *34*, 271–285. [CrossRef]
11. Stalder, A.; Kulik, G.; Sage, D.; Barbieri, L.; Hoffmann, P. A snake-based approach to accurate determination of both contact points and contact angles. *Colloid Surf. A* **2006**, *286*, 92–103. [CrossRef]
12. Cai, D.; Neyer, A.; Kuckuk, R.; Heise, H.M. Raman, mid-infrared, near-infrared and ultraviolet–visible spectroscopy of PDMS silicone rubber for characterization of polymer optical waveguide materials. *J. Mol. Struct.* **2010**, *976*, 274–281. [CrossRef]

13. Tielens, F.; Gervais, C.; Lambert, J.F.; Mauri, F.; Costa, D. Ab initio study of the hydroxylated surface of amorphous silica: A representative model. *Chem. Mater.* **2008**, *20*, 3336–3344. [CrossRef]
14. Grimoldi, E.; Zanini, S.; Siliprandi, R.; Riccardi, C. AFM and contact angle investigation of growth and structure of pp-HMDSO thin films. *Eur. Phys. J. D* **2009**, *54*, 165–172. [CrossRef]
15. Benítez, F.; Martínez, E.; Esteve, J. Improvement of hardness in plasma polymerized hexamethyldisiloxane coatings by silica-like surface modification. *Thin Solid Films* **2000**, *377*, 109–114. [CrossRef]
16. Anderson, R.C.; Muller, R.S.; Tobias, C.W. Chemical surface modification of porous silicon. *J. Electrochem. Soc.* **1993**, *140*, 1393–1396. [CrossRef]
17. Yamada, T.; Lukac, P.J.; Yu, T.; Weiss, R.G. Reversible, room-temperature, chiral ionic liquids. Amidinium carbamates derived from amidines and amino-acid esters with carbon dioxide. *Chem. Mater.* **2007**, *19*, 4761–4768. [CrossRef]
18. Honda, R.Y.; Mota, R.P.; Batocki, R.; Santos, D.; Nicoleti, T.; Kostov, K.; Kayama, M.; Algatti, M.; Cruz, N.; Ruggiero, L. Plasma-polymerized hexamethyldisilazane treated by nitrogen plasma immersion ion implantation technique. *J. Phys. Conf. Ser.* **2009**, *167*, 012055. [CrossRef]
19. Wei, Q.; Haag, R. Universal polymer coatings and their representative biomedical applications. *Mater. Horiz.* **2015**, *2*, 567–577. [CrossRef]

Article

Properties of Post-Consumer Polyethylene Terephthalate Coating Mechanically Deposited on Mild Steels

Elisângela Silva [1,*], Michele Fedel [2], Flavio Deflorian [2], Fernando Cotting [1] and Vanessa Lins [1]

[1] Department of Chemical Engineering, University of Minas Gerais, Av. Pres. Antônio Carlos, 6627, Pampulha, Belo Horizonte 31270-901, Brazil; fernando@deq.ufmg.br (F.C.); vlins@deq.ufmg.br (V.L.)

[2] Department of Industrial Engineering, University of Trento, via Sommarive n. 9, 38123 Trento, Italy; michele.fedel@unitn.it (M.F.); flavio.deflorian@unitn.it (F.D.)

* Correspondence: silvaelisangelax@gmail.com; Tel.: +55-31-3409-1737

Received: 16 October 2018; Accepted: 28 December 2018; Published: 5 January 2019

Abstract: An anticorrosive coating of post-consumer polyethylene terephthalate (PET) was applied on carbon steel by using an industrial press. The PET layer showed a good adhesion on the substrate, evaluated by using a pull off test, when compared with the traditional organic coatings. In addition, scanning electron microscopy (SEM) analysis showed that the PET layer was uniform, homogeneous, and free of cracks or defects. The Fourier-transform infrared spectroscopy (FTIR) and differential scanning calorimetry (DSC) proved that the PET properties were not affected by the deposition process. The PET organic coating is a promising coating, due to its corrosion resistance evaluated by using salt spray tests, even though the applied thickness of 65 μm was considered thin for a high-performance coating. The electrochemical impedance spectroscopy (EIS) showed that the PET coating has a capacitive effect and its electrochemical behavior was not affected as the exposure time increased, resulting in an impedance modulus value of 10^{10} $\Omega \cdot cm^2$, after 576 h of immersion in an aqueous solution of NaCl 3.0 wt %.

Keywords: corrosion; electrochemical impedance spectroscopy (EIS); organic coatings; polyethylene terephthalate (PET)

---

1. Introduction

An alternative contribution to solve the problem of urban solid waste (USW) is to use solid polymeric waste to develop coatings for steels [1]. Carbon steel has excellent mechanical properties and low cost but has some disadvantages such as low wear and corrosion resistance in various media [2]. The most employed method to protect the carbon steel against corrosion process has been the organic coatings. Polymeric layers are applied on the carbon steel to avoid the contact between metallic surface and the aggressive environment. The barrier protection conferred by the polymeric coatings can be associated to their water uptake property [3–10].

The organic coatings based on polyester, epoxy, and polyurethane resins, which present a lower water uptake value compared to alkyd resins, are widely used. Nevertheless, for long exposure time, the absorbed water reaches the metal and the corrosion process is established. In order to retard this process, thick films (from tens of μm to a few mm) of these organic coatings are applied on the substrate, thus causing an increase in the total cost of this project [8–15].

Lins et al. [1] demonstrated that post-consumer commingled polymer (PCCP) coatings on carbon steels can be produced by thermal spraying and investigating their erosion behavior. Literature reports the use of post-consumer tires to develop sustainable polymeric composite materials which are characterized by good mechanical and functional properties [3]. Çinar and Kar [4] mixed PET waste particles in a screwed extruder with marble dust to produce a composite material.

Municipal solid waste management has received a great deal of attention as countries such as India, which produces an estimated quantity of 50–600 million tons of urban solid waste annually [5]. The municipal solid waste is mainly composed of metals, glass and plastics. The municipal solid waste of Bangalore, India, contains 6.23% of plastics [5]. In 2016, 27.1 million tons of plastic waste was collected through official schemes in the EU28 plus Norway and Switzerland to be treated, and for the first time, more plastic waste was recycled than landfilled [6].

One of the main components of plastic fraction of USW is the polyethylene terephthalate (PET) which is one of the most versatile polymers available, which makes it the most produced and used polymer worldwide [7,8]. PET has characteristics such as flexibility, transparency, good adhesion, low permeability to liquids and gases, high thermal resistance and low cost, all of which enable PET to be an excellent candidate for a protective coating with low thickness [8–11]. Literature reports PET deposition on steels by using low-velocity flame spray technology [9].

In this work, post-consumer PET was deposited on carbon steels by using mechanical deposition. Mechanical PET deposition presents simplicity, lower cost in relation to the thermal spraying, and was not found in literature to the best of our knowledge. The process of comminution of PET from the post-consumer bottles until a powder of the desired granulometry is not a simple process; one reason for this relies on the fact that PET is hygroscopic. Based on the obtained results, a procedure was developed for the comminution of PET from the post-consumer bottles, which has proved to be very efficient. The effect of the mechanical deposition on the PET structural and thermal properties was investigated by FT-IR spectroscopy and differential scanning calorimetry (DSC) measurements. The protection properties provided by the PET coatings were evaluated by using electrochemical impedance spectroscopy (EIS) analysis and exposure in the neutral salt spray chamber (NSST, according to the ASTM B117 [12] and ISO 9227-2017 standards [13]). The obtained results have been compared with literature data of pristine paints employed for corrosion protection purposes. In addition, adhesion tests were performed by the pull-off test, in order to characterize the interaction between PET coating and steel substrate.

2. Materials and Methods

2.1. Materials

The post-consumer PET bottles were collected from household waste; the labels and the lid were removed and discarded. The bottles were washed in soapy water. The grinding process allowed a yield of 85% of the overall use of the bottle; only the top and bottom of the bottles were not used.

The substrate used was Q-Panel type R steel with dimensions 80 mm × 40 mm × 1 mm. The composition of this carbon steel was 0.15 C, 0.60 Mn, 0.030 P, 0.035 S (wt %). Prior to deposition, a blast pre-treatment of the steel samples was used, creating a roughness profile on the carbon steel surface (R_z) of approximately 20 µm. An SBC 350 sandblasting machine (Nova, Rehovot, Israel) was used with Garnet filter sand. After this process, the plate was washed with acetone in an ultrasonic bath for 2 min and dried with clean air.

2.2. Deposition Process

The amount of 0.8 g of PET ground with particle size of approximately 0.07 mm was deposited on carbon steel. The set was placed in the press for heating at 260 °C for 5 min without pressure; there after a pressure of 0.5 ton was applied for 2 min. The material remained in the press and was cooled for 5 min. The coated steel was then removed from the press and conditioned at room temperature for 24 h. The PET coating thickness was 65 ± 5 µm.

2.3. Coating Characterization

The morphology of the coating was examined by using scanning electron microscopy (SEM), with a FEG-Quanta 200 FEI equipment (FEI Company, Fremont, CA, USA). The coated steel was also characterized by using Fourier transform infrared spectroscopy employing an attenuated total reflection (ATR) geometry, by means of Varian 4100 Excalibur Series equipment (Santa Clara, CA, USA). The wavelength range was 500–4000 cm^{-1} and the resolution of 4 cm^{-1}. The powder and PET coating are from the same batch before and after pressing in Caver laboratory press, model 2699, Ser. No. 2699-12748, Fred S. carver Inc., Wabash, IN, USA.

The differential scanning calorimetry (DSC) test was performed using the Mettler DSC30 (Mettler Toledo, New York, NY, USA) equipment in three cycles, with heating from 0 to 300 °C under a nitrogen flow of 10 mL·min^{-1}, followed by cooling and heating from 0 to 300 °C, at a heating rate of 10 °C min^{-1}. Thermal analysis provided the glass transition temperature (T_g), crystallization temperature (T_c) and melting temperature (T_m). The crystallinity value (χ_c) was calculated by using the values of the endothermic melting peak using the Equation (1):

$$\chi_c = \frac{\Delta H_m}{\Delta H_m^0} \tag{1}$$

where ΔH_m is melting enthalpy, and ΔH_m^0 the melting enthalpy for polyethylene terephthalate, considered 140 J·g^{-1} [14].

The adhesion of the coating was evaluated by using the pull-off test according to the ASTM D4541 standard [15]. Dollies of 20 mm were glued on the plate coated with Huntsman Araldite 2000 glue; the cure occurred in a period of 24 h. The equipment used was a De Felsko PosiTest AT-M (Ogdensburd, NY, USA). The tests were performed in triplicate.

Coating thickness was measured by using a digital layer thickness gauge on a ferrous base Digi-Derm Mitutoyo (Aurora, CO, USA). Ten measurements were collected on the coated steel.

The size distribution of the PET particles was performed by laser diffraction in the Mastersizer 3000 equipment (Worcestershire, UK). The obtained values were $D(4,3)$, which represents the average size distribution; $D_v(10)$, which represents that only 10% of the particles were smaller than the displayed value; $D_v(50)$, showing that 50% of the particles were larger, and 50% smaller than the determined value; and $D_v(90)$, which reveals that 90% of the particles were below this value.

2.4. Corrosion Resistance Evaluation

Corrosion tests were performed using an Equilam salt spray chamber (Diadema, Brazil) in compliance with the ASTM B117 standard [12] for 500 h. The solution used was 5.0 wt % NaCl with pH between 6.5 and 7.2. A linear scratch was made on the coated steel with a tungsten carbide tool at an angle of 60°.

Electrochemical impedance spectroscopy was performed by using Metrohm AUTOLAB 302 N potentiostat/FRA (Utrecht, The Netherlands) at room temperature. The measurements were performed using a three-electrode electrochemical cell; a platinum foil was used as a counter electrode, a Ag/AgCl$_{(sat)}$ reference electrode was used, and the coated samples were employed as working electrodes. A polyvinylchloride (PVC) tube was affixed with silicone glue on the coated plate, delimiting the exposure area to 6.25 cm^2, where the 3.0 wt % NaCl solution was poured. The open circuit potential was measured for one hour or until stabilization. The EIS measurements were collected at the OCP after 1, 24, 192, 360, 480, and 576 h of immersion. The potential signal amplitude applied was 10 mV (rms), and the frequency range analyzed was 10^5–10^{-2} Hz. The data collected were analyzed by using the ZView2 software.

Evolution of the capacitance of the film was employed to measure the phenomena of water absorption since the presence of moisture modifies the dielectric constant of the polymer [16,17].

The ratio of the dielectric constant (ε) of the coating was calculated using Equation (2) [18]:

$$\varepsilon = \frac{C_c}{\varepsilon_0 A} L \tag{2}$$

ε_0 is the dielectric constant of free space (8.854 $\times$ 10^{-12} F m^{-1}), A is the coating area and L is the coating thickness.

The fraction of water volume of coating at saturation ($\varnothing_\infty$) was calculated by using the Brasher and Kingsbury (BK) formula:

$$\varnothing_\infty = \frac{\log C_\infty / C_0}{\log C_t} \tag{3}$$

where C_∞ is the coating capacitance at immersion time t (determined from EIS data); C_0 is the initial coating capacitance and C_t is the relative permittivity of the water. The parameters C_∞ and $\varnothing$ are dependent on the immersion time [19].

3. Results and Discussion

3.1. Characterization Results

Figure 1 shows the PET powder produced from post-consumer bottles. The particles are heterogeneous in shape and size. More elongated and other equiaxial particles can be identified. Particle sizes have been calculated from measurement of particle areas using the IMAGEJ-win64 software. The PET powder presented a particle area between 430 and 51,200 µm^2, and 90% of PET particles showed the size of 635 µm.

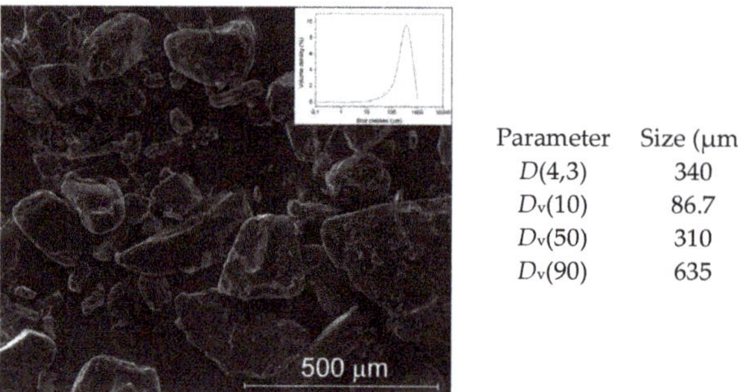

Parameter	Size (µm)
$D(4,3)$	340
$D_v(10)$	86.7
$D_v(50)$	310
$D_v(90)$	635

Figure 1. PET powder granulometric distribution.

Figure 2 presents the cross-sectional section of the coated steel obtained by using SEM analysis. The film is uniform and free of cracks and voids.

Figure 3 shows the FT-IR spectra of PET powder and coating, and Table 1 shows characteristic absorptions of functional groups present in PET powder and PET coating. The PET absorption band at 1715 cm^{-1} is attributed to vibrations of the carbonyl group of saturated esters; the bands at 724 and 871 cm^{-1} are due to the interaction of polar ester groups and benzene rings, vibration =C–H out of the plane. The 1097 cm^{-1} band is associated with the stretching vibration mode of C–O bonds and bands at 1242 cm^{-1} are associated to the specific absorption of the terephthalate group (OOC–C$_6$H$_4$–COO). Bands related to the asymmetric deformation in CH$_2$, at 1407 and 1018 cm^{-1}, were identified [20–22]. At 2960 cm^{-1}, absorption associated to the symmetrical stretch of the C–H bond is observed, with a higher intensity for the PET powder than PET coating. Compared to the coating, the PET powder showed a broad band around 3600 cm^{-1} probably due to the water absorption. Notice that the FT-IR spectra of the powder and of the coating are very similar. The coating deposition treatment does not seem to affect the chemical structure of the polymer.

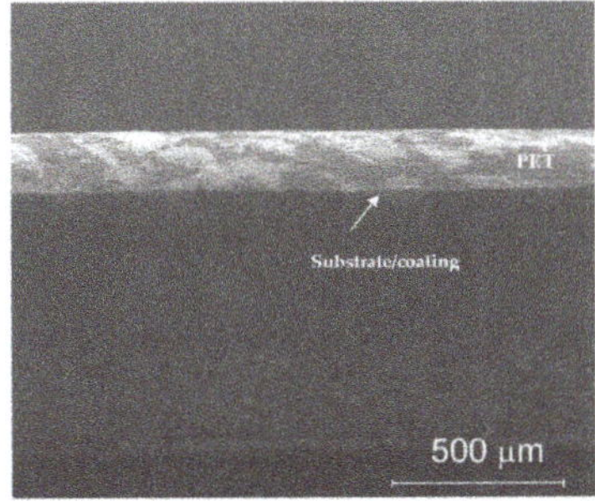

Figure 2. Cross-sectional view of PET coating.

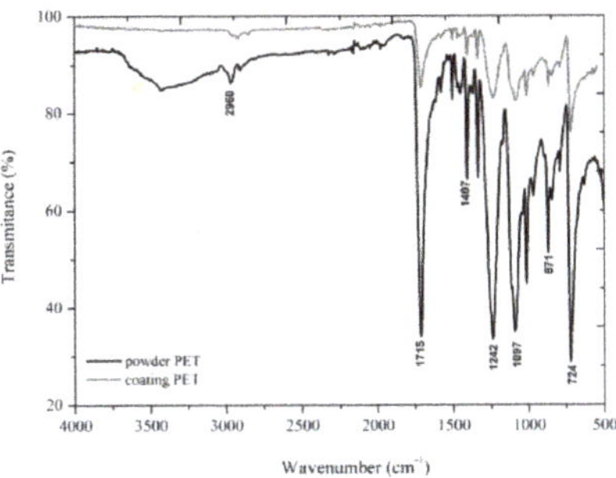

Figure 3. FTIR spectra of PET powder and PET coating.

Table 1. Characteristic absorptions of functional groups present in PET powder and PET coating.

Wavenumber (cm^{-1})	Functional Groups
3200–3600	O–H
2850–3000	C–H aliphatic
1715–1740	stretch C=O unconjugated ketone, ester and carboxylic acid group
1600–1690	stretch C=O aromatic ketone and conjugated aldehydes
1407, 1018, 1465	CH_2–
1270–1097	ester group vibration
1043–972	stretching of ether group C–O–C
871, 724	Vibration =C–H benzene ring off plan

Figure 4 shows DSC results for PET powder and PET coating (removed from the mild steel substrate to carry out the analysis). Table 2 summarizes the thermal analysis results for PET powder and coating. Analyzing the T_g values, the onset values for PET powder and PET coating are very close being 73.8 and 72.8 °C, respectively, within the margin of error of the measurement that is 2 °C. The glass transition temperature is the temperature at which the carbon–carbon bonds become more flexible. The PET coating is rigid at room temperature, since it is below T_g.

The similar T_g values of powder and coating do not indicate significant damage in the reorganization of the polymer chain of the coating during deposition. As far as thermally sprayed PET on mild steels are concerned, Duarte et al. [9] reported that the T_g of coating was lower than that of the powder, which indicated a thermal degradation of the PET during the thermal deposition, decreasing its molar mass. In [9], the T_g value for the PET powder was 79 °C, about 5 °C above the value obtained in this work. The glass transition temperature depends on the heating rate which is the same in this work and in reference [8]. In the case of the thermal spray technique, the polymer is subjected to high temperatures above 1000 °C, above the degradation temperature of the PET which is about 450 °C. Demirel [23] studied mound surface temperature in injection stretch blow molding (ISBM) of polyethylene terephthalate (PET) bottles for carbonated soft drinks (CSD) storage. The cited author reported the effect of mound surface temperature on T_g of PET bottles which varied between 58.3 and 59.4 °C.

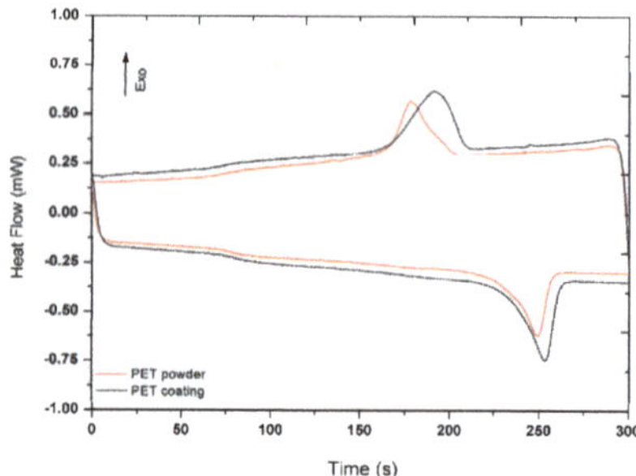

Figure 4. DSC results of PET powder and PET coating on mild steel.

Table 2. Glass transition and melting temperatures, and crystallinity of PET powder and PET coatings.

Sample	Glass Transition Temperature (T_g) (°C)		Crystallization Temperature (T_c) (°C)	ΔH_c (J g^{-1})	Melting Temperature (T_m) (°C)	ΔH_m (J g^{-1})	χ_c (%)	Decomposition Temperature (T_d) (°C)
	Onset	Midpoint						
PET powder	73.8	78.6	195.0	35.6	233.0	35.7	25.5	454.7
PET coating	72.8	80.0	209.0	46.8	235.0	44.4	31.7	455.0

The crystallization temperature of the coating showed an increase of 6.7% in relation to the powder, as seen in Figure 4 and Table 1. This phenomenon may have occurred due to the PET melting at 260 °C for application on the steel, and due to the slow cooling process, which provides time to recrystallization of part of the material. The χ_c values also confirm the crystallinity increase, being 25.5% for the powder and 31.7% for the coating. Takeshita et al. [24] studied the influence of cooling time on the physical properties of the polyester powder coating. They concluded that the cooling time influences the crystallinity values of the polymer and that the ratio between the cooling time and the degree of crystallinity is almost linear. In this sense, the increase in crystallinity of the PET coating was due to the cooling process of the coating deposition. Demirel [23] showed values of crystallinity of PET bottles similar to that found in this work, in the range from 21% up to 29%. Melting temperatures for PET powder and coating were similar (233 and 235 °C, respectively). Literature reports values of PET melting temperature in the range of 242–260 °C [22,25].

PET degradation occurred at temperatures of 454.7 and 455.0 °C for powder and coating, respectively. As the PET reached 260 °C in the deposition process, degradation did not occur. Duarte et al. [9] reported degradation temperatures of 444 °C for PET powder and 437–446 °C for thermally sprayed PET coating.

Dry adhesion test was carried out by means of pull-off test. The detachment of the dolly occurred at the interface between the glue and the coating, at about 5 MPa. Based on these findings, we can assume that the adhesion strength is greater than 5 MPa. The value found for PET coating was compared with literature values for polyester powder coatings. As a result, we can observe in Table 3 that the adhesion values for the polyester family is very close to the value found in this work.

Table 3. Thickness and adhesion for polyester and polyethylene terephtahalate.

Samples	Thickness (µm)	Adhesion (MPa)	Reference
Polyester	291–294	4.8–6.2	[24]
Carboxylate polyester resin (URALAC)	50.0 ± 5	1.7–3.0	[26]
Saturated carboxylated polyester resin	90.0 ± 10	4–6.8	[27]
Post-consumer polyethylene terephthalate	65 ± 5	>5	This work

3.2. Corrosion Resistance of PET Coated Steel

3.2.1. Results of Exposure of the Coating on Scratched Samples after 480 h in Salt Spray Chamber

Salt spray results are shown in Figure 5 for up to 480 h of exposure in chamber.

The salt spray tests showed a reduced extent of coating delamination resulting from the scratch. No blisters where observed during the 480 h of immersion. As the test time increased, a marked increase in the corrosion of the carbon steel in the scratch area was observed. However, the cathodic front did not advance underneath the coating. No coating detachment was observed during exposure time of up to 480 h.

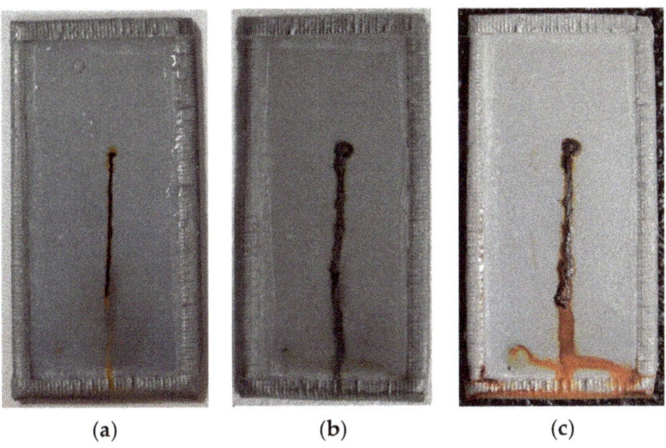

Figure 5. Photographs of PET coated steel after exposure for (**a**) 72, (**b**) 240, and (**c**) 480 h in a salt spray chamber.

3.2.2. Electrochemical Impedance Spectroscopy (EIS)

Figure 6 shows Bode diagram for PET coated steel in an aqueous solution of NaCl for immersion up to 576 h. The phase angle curves are sometimes scattered, in particular between 10 and 100 Hz and around 0.1 Hz, probably due to the relatively high impedance of the coating during immersion time. According to the impedance modulus diagrams (Figure 6a) the impedance values in the low frequency range (at about 0.01 Hz, $|Z|_{0.01}$) was high and almost stable as it ranged from 1×10^{10} to 3×10^{10} Ω cm^2 throughout immersion time. A straight line with the slope close to -1 was observed in the Bode diagram, thus suggesting an almost capacitive behavior over a wide range of frequencies. A single time relaxation process was observed in the middle frequency range.

After 360 h of immersion, a decrease in impedance in the low frequency domain was observed, however, there is a trend for stabilization (Figure 7). The low frequency impedance was in the order of 1×10^{10}–3×10^{10} Ω cm^2 during all immersion time, thus suggesting that the PET applied coating provided steel corrosion protection. The PET coating seems to be able to retard the direct contact between the corrosive medium and metal substrate. To better investigate the properties of the recycled PET coating, EIS data were fitted employing the equivalent circuit shown in Figure 7. According to [28,29], a $R_e(Q_cR_p)$ was employed to fit the experimental spectra. R_e is the electrolyte resistance. The constant phase element, Q_c, and the resistance, R_p, where attributed to the dielectric properties and the pore resistance of the PET coating, respectively.

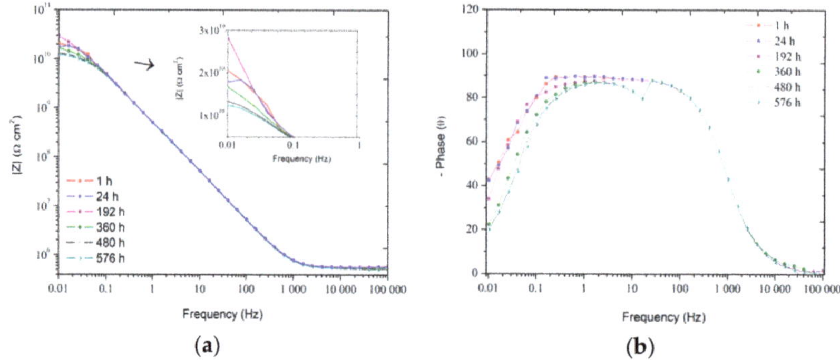

(a)

(b)

Figure 6. (a) Bode diagram of impedance modulus versus frequency for the PET coated steel in a saline environment for 1, 24, 192, 360, 480, and 576 h, with amplification of low frequency region; (b) Bode diagram of phase angle versus frequency for the PET coated steel in a saline environment for 1, 24, 192, 360, 480, and 576 h.

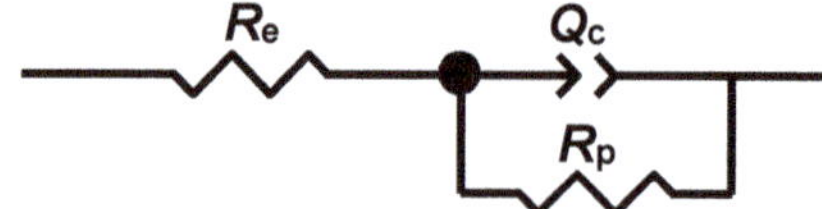

Figure 7. Equivalent circuit for EIS data of PET coated steel in NaCl solution.

Considering the results reported in Table 4, as the values of the '*n*' exponent is about 0.98 during the whole immersion time, the pre-factor of the CPE named Q_c can be assumed as the coating capacitance. For this reason, from now on, the pre-exponential factors of the CPE will be considered throughout the paper as an acceptable approximation of the capacitance values during immersion time. Figure 8 shows the evolution of the normalized coating capacitance (respect to the initial capacitance value, $C_{t=x}/C_{t=0}$) during the first 24 h of immersion. Initially, a sharp increase in the normalized capacitance was observed, probably due to micropores formed in the coating which facilitated water permeation. After the first hours of immersion, the capacitance reached a sort of saturation, remaining stable at a value 1.12/1.14 times the initial capacitance. As far as the pore resistance is concerned, from Table 4 it is possible to observe that it decreased from 1.82×10^{10} to 0.81×10^{10} Ω cm^2 after 576 h due to continuous immersion in the electrolyte, thus suggesting that the recycled PET coatings have excellent barrier properties.

In fact, the pore resistance R_p values of the investigated coatings are comparable or, in particular cases, highly compared to previously reported literature data (Table 5) related to polyester-based paints (derived from pristine materials).

Table 4. Electrochemical parameters for PET coated steel in a saline medium.

Time (h)	$R_p\ 10^{10}$ (Ω cm^2)	Error (%)	Pre-Factor of $Q_c\ 10^{-11}$ (S s^n cm^{-2})	Error (%)	n
1	1.82	5.2	5.30	1.11	0.98
24	1.78	3.77	5.25	1.35	0.98
192	2.10	2.10	5.14	0.72	0.98
360	1.09	1.69	5.25	0.83	0.98
480	0.86	1.62	5.30	0.89	0.98
576	0.81	1.62	5.34	0.90	0.98

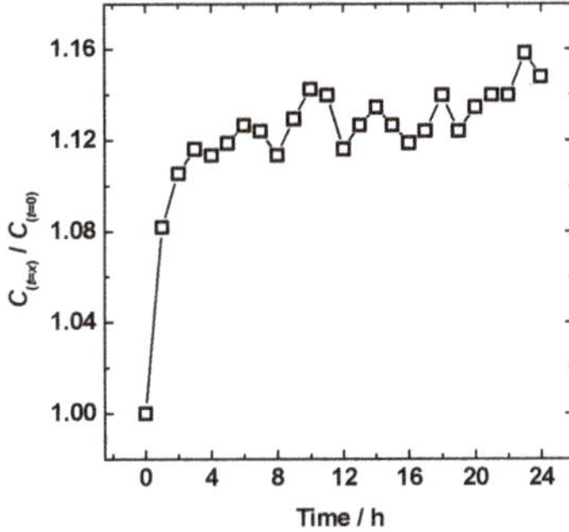

Figure 8. Capacitance versus immersion time of PET coated steel.

Table 5. R_p values for polyester from the literature in electrolytes of 3.0–3.5 wt % of NaCl.

Samples	Thickness (µm)	Time (h)	R_p (Ω cm^2)	Reference
Polyester powder coating	90 ± 10	0–500	$\approx 10^8 \rightarrow 10^5$	[27]
Polyester powder coating	30 ± 2	24–1368	$\approx 10^{10} \rightarrow 10^7$	[28]
Polyester powder coating	45 ± 3	168–1200	$\approx 10^7 \rightarrow 10^4$	[30]
Polyester/epoxy powder coating	45 ± 3	168–1200	$\approx 10^8 \rightarrow 10^5$	[30]
Polyester resin	30–40	0–720	$\approx 10^7 \rightarrow 10^4$	[31]

Relative permittivity of dry coating ε_d and fraction of water volume $\varnothing$ were calculated by Equations (2) and (3). The value of ε_d for PET coating was 5.0: this value is in accordance with relative permittivity reported in the literature to dry coating 3–8 [29,32,33]. The coating to be considered protective and water resistant must show the value of $\varnothing$ between 2.0% and 15.98%, approximately [29,34]. The value of fraction of water volume estimated for the PET coating was 3.2% after 24 h of immersion in 3.0 wt % NaCl solution.

4. Conclusions

The DSC results indicated no significant damage to the polymer produced by the mechanical deposition (press recycled).

The PET organic coating with a thin layer of 65 µm presented good adhesion to the substrate, superior to 5 MPa, evaluated by using the pull-off test, as well as a high corrosion protection for carbon steel during long immersion times in the NaCl 3.0 wt % electrolyte; both of which are important characteristics for an anticorrosive organic coating. The polarization resistance of PET coated steel in a saline solution was 8.1×10^9 Ω cm^2 after 576 h of immersion. The value of fraction of water volume absorbed was 3.2% indicating the excellent protective action of coating.

The PET organic coatings can be considered as an alternative, both for the PET recycling and for a new anticorrosive coating.

Wear evaluation of PET coating will be performed in further investigation.

Author Contributions: Conceptualization, E.S., V.L., F.D. and M.F.; Methodology, E.S.; Validation, E.S., F.D., M.F. and F.C.; Formal Analysis, E.S., M.F. and F.C.; Investigation, E.S. and M.F.; Writing–Original Draft Preparation, E.S. and V.L.; Writing–Review & Editing, E.S., M.F., F.D., F.C. and V.L.; Supervision, E.S.; Project Administration, E.S., F.D. and M.F.

Funding: This research was funded by Foundation CAPES (Coordination for higher Education Staff Development) (No. 88881.135023/2016-01).

Acknowledgments: The authors would like to sincerely thank the polymers laboratory of University of Trento.

Conflicts of Interest: The authors declare no conflict of interest.

References

1. Lins, V.F.C.; Branco, J.R.T.; Diniz, F.R.C.; Brogan, J.C.; Berndt, C.C. Erosion behavior of thermal sprayed, recycled polymer and ethylene–methacrylic acid composite coatings. *Wear* **2007**, *262*, 274–281. [CrossRef]
2. Le, L.; Mojtaba, M.; Chun, Q.L.; Dilan, R. Effect of corrosion and hydrogen embrittlement on microstructure and mechanical properties of mild steel. *Constr. Build. Mater.* **2018**, *170*, 78–90. [CrossRef]
3. Sienkiewicz, M.; Janik, H.; Borzędowska-Labuda, K.; Kucińska-Lipka, J. Environmentally friendly polymer-rubber composites obtained from waste tyres: A review. *J. Clean. Prod.* **2017**, *147*, 560–571. [CrossRef]
4. Çınar, M.E.; Kar, F. Characterization of composite produced from waste PET and marble dust. *Constr. Build. Mater.* **2018**, *163*, 734–741. [CrossRef]
5. Ramachandra, T.V.; Bharath, H.A.; Kulkarni, G.; Han, S.S. Municipal solid waste: Generation, composition and GHG emissions in Bangalore, India Review article. *Renew. Sustain. Energy Rev.* **2018**, *82*, 1122–1136. [CrossRef]
6. Plastics Europe. Available online: https://www.plasticseurope.org/application/files/5715/1717/4180/Plastics_the_facts_2017_FINAL_for_website_one_page.pdf (accessed on 9 August 2018).
7. Shen, L.; Worrell, E.; Patel, M.K. Open-loop recycling: A LCA case study of PET bottle-to-fibre recycling. *Resour. Conserv. Recycl.* **2010**, *55*, 34–52. [CrossRef]
8. Awaja, F.; Pavel, D. Recycling of PET. *Eur. Polym. J.* **2005**, *41*, 1453–1477. [CrossRef]
9. Duarte, L.T.; Paula, E.M.; Branco, J.R.T.; Lins, V.F.C. Production and characterization of thermally sprayed polyethylene terephthalate coatings. *Surf. Coat. Technol.* **2004**, *182*, 261–267. [CrossRef]
10. Phetphaisit, C.W.; Namahoot, J.; Saengkiettiyut, K.; Ruamcharoen, J.; Ruamcharoen, P. Green metal organic coating from recycled PETs and modified natural rubber for the automobile industry. *Prog. Org. Coat.* **2015**, *86*, 181–189. [CrossRef]
11. Sanaee, Z.; Mohajerzadeh, S.; Zand, K.; Gard, F.S.; Pajouhi, H. Minimizing permeability of PET substrates using oxygen plasma treatment. *Appl. Surf. Sci.* **2011**, *257*, 2218–2225. [CrossRef]
12. *ASTM B117 Standard Practice for Standard Practice for Operating Salt Spray (Fog) Apparatus*; ASTM International: West Conshohocken, PA, USA, 2012.
13. *ISO 9227-2017 Corrosion Tests In Artificial Atmospheres—Salt Spray Tests*; ISO: Geneva, Switzerland, 2017.
14. Jayakannan, M.; Ramakrishnan, S. Effect of branching on the crystallization kinetics of poly(ethylene terephthalate). *J. Appl. Polym. Sci.* **1999**, *74*, 59–66. [CrossRef]
15. *ASTM D4541 Standard Test Method for Pull-Off Strength of Coatings Using Portable Adhesion Testers*; ASTM International: West Conshohocken, PA, USA, 2017.
16. Deflorian, F.; Fedrizz, L.; Rossi, S.; Bonora, P.L. Organic coating capacitance measurement by EIS: Ideal and actual trends. *Electrochim. Acta* **1999**, *44*, 4243–4249. [CrossRef]
17. Bacon, R.C.; Smith, J.J.; Rugg, F.M. Electrolytic resistance in evaluating protective merit of coatings on metals. *Ind. Eng. Chem.* **1948**, *40*, 161–168. [CrossRef]
18. Zhang, J.; Hu, J.; Zhang, J.; Cao, C. Studies of impedance models and water transport behaviors of polypropylene coated metals in NaCl solution. *Prog. Org. Coat.* **2004**, *49*, 293–301. [CrossRef]
19. Wang, H.; Zhou, Q. Evaluation and failure analysis of linseed oil encapsulated self healing anticorrosive coating. *Prog. Org. Coat.* **2018**, *118*, 108–115. [CrossRef]
20. Chen, Z.; Hay, J.N.; Jenkins, M.J. FTIR spectroscopic analysis of poly(ethylene terephthalate) on crystallization. *Eur. Polym. J.* **2012**, *48*, 1586–1610. [CrossRef]

21. Chen, Z.; Hay, J.N.; Jenkins, M.J. The thermal analysis of poly(ethylene terephthalate) by FTIR spectroscopy. *Thermochim. Acta* **2013**, *552*, 123–130. [CrossRef]
22. Parvinzadeh, M.; Moradian, S.; Rashidi, A.; Yazdanshenas, M.E. Surface characterization of polyethylene terephthalate/silica nanocomposites. *Appl. Surf. Sci.* **2010**, *256*, 2792–2802. [CrossRef]
23. Demirel, B. Optimisation of mould surface temperature and bottle residence time in mould for the carbonated soft drink PET containers. *Polym. Test* **2017**, *60*, 220–228. [CrossRef]
24. Takeshitaa, Y.; Sawadaa, T.; Handaa, T.; Watanuki, Y.; Kudo, T. Influence of air-cooling time on physical properties of thermoplastic polyester powder coatings. *Prog. Org. Coat.* **2012**, *75*, 584–589. [CrossRef]
25. Guçlu, G.; Orbay, M. Alkyd resins synthesized from postconsumer PET bottles. *Prog. Org. Coat.* **2009**, *65*, 362–365. [CrossRef]
26. Mirabedini, S.M.; Kiamaneshb, A.A. The effect of micro and nano-sized particles on mechanical and adhesion properties of a clear polyester powder coating. *Prog. Org. Coat.* **2013**, *76*, 1625–1632. [CrossRef]
27. Puiga, M.; Cabedoa, L.; Graceneaa, J.J.; Jiménez-Morales, A.; Gámez-Pérez, J.; Suay, J.J. Adhesion enhancement of powder coatings on galvanised steel by addition of organo-modified silica particles. *Prog. Org. Coat.* **2014**, *77*, 1309–1315. [CrossRef]
28. Jegdić, B.V.; Bajat, J.B.; Popić, J.P.; Stevanović, S.I.; Mišković-Stanković, V.B. The EIS investigation of powder polyester coatings on phosphated low carbon steel: The effect of NaNO$_2$ in the phosphating bath. *Corros. Sci.* **2011**, *53*, 2872–2880. [CrossRef]
29. Ding, R.; Jiang, J.; Gui, T. Study of impedance model and water transport behavior of modified solvent-free epoxy anticorrosion coating by EIS. *J. Coat. Technol. Res.* **2016**, *13*, 501–515. [CrossRef]
30. Mafi, R.; Mirabedini, S.M.; Naderi, R.; Attar, M.M. Effect of curing characterization on the corrosion performance of polyester and polyester/epoxy powder coatings. *Corros. Sci.* **2008**, *50*, 3280–3286. [CrossRef]
31. Ismail, L.; Ramesh, K.; Mat Nor, N.A.; Jamari, S.K.M.; Vengadaesvaran, B.; Arof, A.K. Performance of polyester/epoxy binder coating system—Studies on coating resistance, adhesion and differential scanning calorimetry. *Pigment Resin Technol.* **2016**, *45*, 158–163. [CrossRef]
32. Moreno, C.C.; Hernández, S.; Santana, J.J.; González-Guzmán, J.; Souto, R.M.; González, S. Characterization of water uptake by organic coatings used for the corrosion protection of steel as determined from capacitance measurements. *Int. J. Electrochem. Sci.* **2012**, *7*, 7390–7403.
33. Shreepathi, S.; Naik, S.M.; Vattipalli, M.R. Water transportation through organic coatings: Correlation between electrochemical impedance measurements, gravimetry, and water vapor permeability. *J. Coat. Technol. Res.* **2012**, *9*, 411–422. [CrossRef]
34. Fredj, N.; Cohendoz, S.; Mallarino, S.; Feaugas, X.; Touzain, S. Evidencing antagonist effects of water uptake and leaching processes in marine organic coatings by gravimetry and EIS. *Prog. Org. Coat.* **2010**, *67*, 287–295. [CrossRef]

Article

Studies on Synthesis and Characterization of Aqueous Hybrid Silicone-Acrylic and Acrylic-Silicone Dispersions and Coatings. Part I

Janusz Kozakiewicz [1,*], Joanna Trzaskowska [1], Wojciech Domanowski [1], Anna Kieplin [2], Izabela Ofat-Kawalec [1], Jarosław Przybylski [1], Monika Woźniak [2], Dariusz Witwicki [2,†] and Krystyna Sylwestrzak [1]

[1] Department of Polymer Technology and Processing, Industrial Chemistry Research Institute, Rydygiera 8, 01-793 Warsaw, Poland; joanna.trzaskowska@ichp.pl (J.T.); wojciech.domanowski@ichp.pl (W.D.); izabela.ofat-kawalec@ichp.pl (I.O.-K.); jaroslaw.przybylski@ichp.pl (J.P.); krystyna.sylwestrzak@ichp.pl (K.S.)

[2] D&R Dispersions and Resins Sp. z o. o., Duninowska 9, 87-800 Włocławek, Poland; anna.kieplin@d-resins.com (A.K.); monika.wozniak@d-resins.com (M.W.)

* Correspondence: janusz.kozakiewicz@ichp.pl; Tel.: +48-500-433-297

† The author passed away (1963–2018).

Received: 15 November 2018; Accepted: 23 December 2018; Published: 3 January 2019

Abstract: The objective of the study was to investigate the effect of the method of synthesis on properties of aqueous hybrid silicone-acrylic (SIL-ACR) and acrylic-silicone (ACR-SIL) dispersions. SIL-ACR dispersions were obtained by emulsion polymerization of mixtures of acrylic and styrene monomers (butyl acrylate, styrene, acrylic acid and methacrylamide) of two different compositions in aqueous dispersions of silicone resins synthesized from mixtures of silicone monomers (octamethylcyclotetrasiloxane, vinyltriethoxysilane and methyltriethoxysilane) of two different compositions. ACR-SIL dispersions were obtained by emulsion polymerization of mixtures of the same silicone monomers in aqueous dispersions of acrylic/styrene copolymers synthesized from the same mixtures of acrylic and styrene monomers, so the compositions of ACR and SIL parts in corresponding ACR-SIL and SIL-ACR hybrid dispersions were the same. Examination of the properties of hybrid dispersions (particle size, particle structure, minimum film forming temperature, T_g of dispersion solids) as well as of corresponding coatings (contact angle, water resistance, water vapour permeability, impact resistance, elasticity) and films (tensile strength, elongation at break, % swell in toluene), revealed that they depended on the method of dispersion synthesis that led to different dispersion particle structures and on composition of ACR and SIL part. Generally, coatings produced from hybrid dispersions showed much better properties than coatings made from starting acrylic/styrene copolymer dispersions.

Keywords: aqueous polymer dispersions; silicone-acrylic; acrylic-silicone; hybrid particle structure; coatings

---

1. Introduction

Aqueous polymer dispersions are currently produced in quantities exceeding globally 20 million tons per annum [1] and are commonly used, inter alia, as binders for organic coatings, especially for aqueous dispersion-based architectural paints. The reason for a great increase in research and business interest in that specific group of products is not only the fact that they are environmentally friendly, but also the possibility of tailoring the composition and structure of dispersion particle in order to achieve desired performance characteristics of the final coating. If the particles have a hybrid (It is worth to note that generally a "hybrid material" is a material that is composed of at least two components mixed at molecular scale [2] and although this term is normally used for polymer-inorganic structure composites [3], it can be also applied to polymer-polymer systems) structure (i.e., are composed of at least two different polymers) and their diameter is less than 100 nm they may be called "dispersion nanoparticles with hybrid structure" within which the occurrence of specific interactions between these polymers optionally leading to synergistic effect may be expected. Then, due to a synergistic effect, new and sometimes quite unexpected features of coatings or films made using such hybrid dispersions as binders may be found—see Figure 1.

(a)   (b)

Figure 1. Differences between mixture of two aqueous dispersions of different polymers (polymer 1—blue color and polymer 2—red color) and aqueous dispersion with hybrid particle structure composed of the same two different polymers. In the mixture of two dispersions (**a**) synergistic effect is much less probable than in dispersion with hybrid particle structure (**b**) [4].

Although some authors reported that certain specific properties of coatings could be improved by using blends of dispersions of different polymers (e.g., dirt resistance could be enhanced this way [5]), clear advantages of application of dispersions with hybrid particle structure as coating binders have been confirmed in the literature, e.g., [1,4,6] and descriptions of methods applied for synthesis of such dispersions can be found in books and review papers, e.g., [7–10]. The particle morphology that is most frequently referred to in the research articles is a "core-shell" structure, but other morphologies, like core-double shell, gradient, eye-ball-like, raspberry-like, fruit cake or embedded sphere can also be obtained [11,12]. It has been proved [8] that not only the hybrid dispersion particle size and chemical composition, but also its morphology can significantly affect the properties of both dispersions and coatings produced from such dispersions. Therefore, investigation of the hybrid dispersion systems is of great interest to many researchers.

According to [7], the following general approaches can be applied to synthesis of hybrid aqueous dispersions in the emulsion polymerization process:

- A process where monomer X is polymerized in aqueous dispersion of polymer Y or monomer Y is polymerized in aqueous dispersion of polymer X;
- A process where monomer X is added to aqueous dispersion of polymer Y or monomer Y is added to aqueous dispersion of polymer X and left for some time in order to achieve swelling of dispersion particles with the monomer, and only then is polymerization conducted;
- A process where a mixture of monomers X and Y is placed in the reactor before start of polymerization or is added dropwise during the polymerization. However, in this case formation of particles with hybrid structure would be possible only if the corresponding homopolymers are not compatible or either reactivities of monomers or their polymerization mechanisms differ significantly.

As is clear from the literature, e.g., [4,13] the aforementioned methods of hybrid aqueous dispersion synthesis can be successfully applied to obtain different dispersion particle morphologies, depending on the selection of the starting materials (polymers, monomers, surfactants, initiators etc.). It can be expected that if methods (1) or (2) are applied, the "core-shell" morphology will be the most probable one supposing that certain conditions are fulfilled: "core" polymer should be more hydrophobic than "shell" polymer and formation of separate particles of polymer X in the course of polymerization in dispersion of polymer Y resulting from homolytic nucleation is retarded, e.g., by diminishing the monomer and surfactant concentration in the reaction mixture. The mechanism of hybrid particle formation in the emulsion polymerization process has been described in detail, e.g., in [4,13–15] and factors that determine creation of specific particle morphology have been identified. A review of fundamental theoretical aspects of the formation of dispersion particles with a hybrid structure has been published [16].

One of the hybrid aqueous dispersion systems that is most interesting from the point of view of practical application, especially as coating binders, is dispersion with particles containing organic polymer (usually polyacrylate) and silicone. This is because silicone-containing polymer systems allow for achieving specific features of coating surface like e.g., water repellency without affecting its general performance [17]. A comprehensive review of synthesis and characterization of such hybrid silicone-acrylic dispersions as well as of coatings and films or powders produced from them has been published in 2015 [11]. It has been proved in a number of research papers both referred to in that review paper and published later that the unique properties of coatings like e.g., high surface hydrophobicity and water resistance combined with enhanced water vapour permeability and good mechanical properties can be achieved by applying methods (1) to (3) described above to synthesize hybrid dispersions containing silicones where monomer X is acrylic monomer and monomer Y is silicone monomer—see e.g., [18–21] for method (1), [22–24] for method (2) or [24–26] for method (3). If fluoroacrylic monomer was used as monomer X [27–30], the surface hydrophobicity of coatings could be improved even more. It was also proved in our earlier studies [31–33] that the application of method (1) to emulsion polymerization of methyl methacrylate in aqueous silicone resin dispersions led to stable "core-shell" silicone-poly(methyl methacrylate) hybrid dispersions which, after drying, produced corresponding "nanopowders" that were later used as very effective impact modifiers for powder coatings.

In the present study we investigated the effect of the approach to synthesis (method (1) or method (2) as defined above) on the properties of hybrid aqueous dispersions and corresponding coatings. Two different silicone resin dispersions and two different acrylic/styrene copolymer dispersions were used as starting media in which emulsion polymerization of acrylic and styrene monomers or silicone monomers respectively was conducted. The mass ratio equal to 1/3 of silicone part (SIL 1 or SIL 2) to acrylic/styrene part (ACR A or ACR B) in the synthesis was selected based on the assumption supported by the literature [11] that this proportion would be sufficient to observe the influence of the presence of silicone in the dispersion particle on the properties of hybrid dispersions and coatings. Further studies on the effect of SIL/ACR ratio on the properties of hybrid dispersions and coatings are ongoing and will be published soon.

2. Materials and Methods

2.1. Starting Materials

Octamethylcyclotetrasiloxane (D4) was obtained from Momentive (Waterford, NY, USA). Other silicone monomers: vinyltriethoxysilane (VTES) and methyltriethoxysilane (MTES)) were obtained from Evonik (Essen, Germany) under trade names Dynasylan® VTEO and Dynasylan® MTES. Surfactants dodecylbenzenesulphonic acid (DBSA) and Rokanol T18 (nonionic surfactant based on ethoxylated C16–C18 alcohols) were obtained from PCC Exol (Brzeg Dolny, Poland). Emulgator E30 (anionic surfactant based on C15 alkylsulfonate) was obtained from LeunaTenside GmbH (Leuna, Germany). Other standard ingredients used in the synthesis of dispersions (sodium acetate, sodium hydrocarbonate, potassium persulphate and aqueous ammonia solution) were obtained from Standard Lublin (Poland) as pure reagents. Biocide used to protect dispersions from infestation was Acticide MBS obtained from THOR (Wincham, UK). Starting acrylic/styrene copolymer dispersions (ACR A and ACR B) characterized by different T_{gs} were supplied by Dispersions & Resins (D&R, Włocławek, Poland). Monomers applied in synthesis of ACR A and ACR B dispersions by D&R were butyl acrylate (BA) obtained from ECEM, Arkema, Indianapolis, IN, USA, styrene (ST) obtained from KH Chemicals, Helm, Zwijndrecht, The Netherlands, acrylic acid (AA) obtained from Prochema, BASF, Wien, Austria, and methacrylamide (MA) obtained from ECEM, Arkema. Acrylic and styrene monomers were used as received as mixtures designated as A and B with compositions corresponding to compositions of monomers applied to synthesize dispersions ACR A and ACR B, respectively. Exact compositions could not be revealed due to commercial secret, but were appropriately designed to get T_g of dispersion solids at a level of ca. +15 °C (ACR A) and of ca. +30 °C (ACR B). For structures of acrylic monomers-see Figure 2.

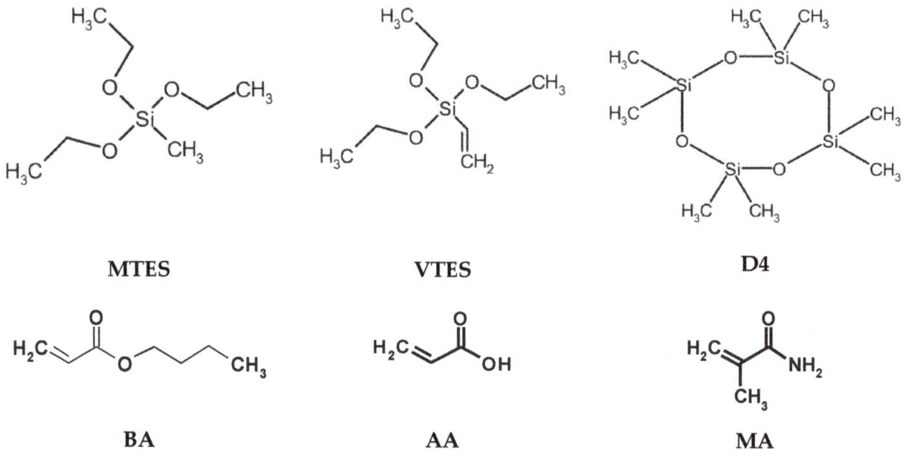

MTES   **VTES**   **D4**

BA   **AA**   **MA**

Figure 2. Structures of silicone monomers used in synthesis of silicone resin dispersions SIL 1 and SIL 2 and acrylic/styrene polymer dispersions ACR A and ACR B.

ACR A and ACR B dispersions were not neutralized after polymerization had been completed in order to ensure the low pH value (ca. 3) that was needed to conduct polymerization of silicone monomers in the process of synthesis of hybrid acrylic-silicone dispersions.

2.2. Synthesis of Silicone Resin Dispersions and Hybrid Silicone-Acrylic and Acrylic-Silicone Dispersions

Silicone resin dispersions (SIL 1 and SIL 2) were synthesized according to the procedure described in [31], although different functional silanes were used along with D4 as silicone monomers—see Figure 2 for the structures of these silicone monomers.

The compositions (wt.%) of mixtures of silicone monomers used in synthesis of SIL 1 and SIL 2 were as follows:

- Mixture designated as 1: D4—84.0%, MTES—9.5%, VTES—6.5%
- Mixture designated as 2: D4—88.0%, VTES—12%

DBSA was used as surfactant and D4 ring-opening catalyst. The reaction that proceeded in the process of SIL 1 and SIL 2 synthesis was simultaneous hydrolysis of trifunctional ethoxysilanes and breaking of Si-O bond in D4 leading to the formation of partially crosslinked poly(dimethylsiloxane) containing unsaturated bonds originating from VTES (see Figure 3).

After distillation of ethanol under vacuum no free VTES or MTES were detected by gas chromatography (GC) in the resulting SIL dispersions, although small amounts of D4 (ca. 0.8%) and ethanol (ca. 0.2%) were still present.

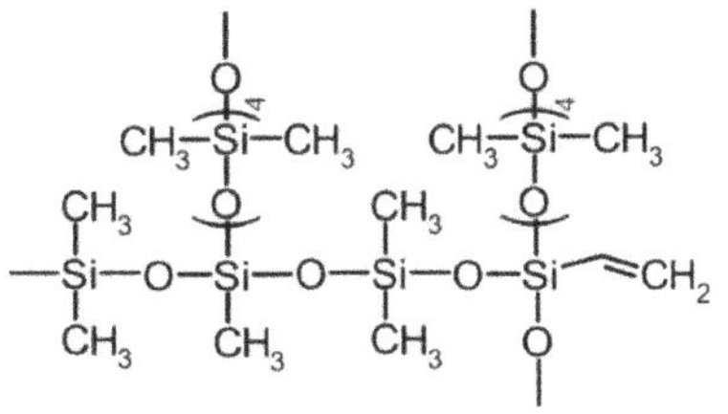

Figure 3. Simplified structure of partially crosslinked silicone resin obtained in synthesis of SIL 1 and SIL 2 dispersions.

Hybrid silicone-acrylic dispersions SIL-ACR 1-A and SIL-ACR 1-B were synthesized by emulsion polymerization of mixtures of acrylic and styrene monomers A and B, respectively, in silicone resin dispersion SIL 1. Hybrid silicone-acrylic dispersions SIL-ACR 2-A and SIL-ACR 2-B were synthesized by emulsion polymerization of mixtures of acrylic and styrene monomers A and B, respectively, in silicone resin dispersion SIL 2. Compositions of acrylic and styrene monomers mixtures A and B corresponded to compositions of acrylic and styrene monomers in starting acrylic/styrene copolymer dispersions ACR A and ACR B. Polymerization was carried out at 78–79 °C for 5 h. After cooling to room temperature, dispersions were neutralized with 25% aqueous NH_3 solution to reach pH ca. 6.0–6.5, then 0.15% of biocide was added and dispersions were filtered through 190 mesh net. Free acrylic and styrene monomers contents as tested by GC were <0.01%. No free VTES or MTES were detected by GC, although small amounts of D4 (ca. 0.4%) and ethanol (ca. 0.1%) were still present. Hybrid acrylic-silicone dispersions ACR-SIL A-1 and ACR-SIL A-2 were synthesized by emulsion polymerization of mixtures of silicone monomers 1 and 2, respectively, in acrylic/styrene copolymer dispersion ACR A. Hybrid acrylic-silicone dispersions (ACR-SIL B-1 and ACR-SIL B-2 were synthesized by emulsion polymerization of mixtures of silicone monomers 1 and 2, respectively, in acrylic/styrene copolymer dispersion ACR B. Compositions of silicone monomers mixtures 1 and 2 corresponded to compositions of silicone monomers in starting silicone resin dispersions SIL 1 and SIL 2. Polymerization was carried out at 88–89 °C for 4 h and then ethanol that was formed in hydrolysis of VTES and MTES was distilled off under vacuum for 3 h. After cooling to room temperature, dispersions were neutralized with 25% NH_3 solution to reach pH ca. 6.0–6.5, then 0.15% of biocide was added and dispersions were filtered through 190 mesh net. No free VTES or MTES or acrylic and styrene monomers were detected by GC in the resulting dispersions, though small amounts of D4 (ca. 0.4%) and ethanol (ca. 0.1%) were still present.

It was essential that the composition and concentration of surfactants remained the same in SIL-ACR and ACR-SIL dispersions, so their properties (and properties of coatings obtained from them) could be compared.

2.3. Characterization of Dispersions

All dispersions were characterized by:

- Solids content, wt.%–percentage of sample mass remaining after drying for 1 h at 80 °C followed by 4 h at 125 °C. The measurements were conducted three times and the mean value was taken.
- pH—using standard indicator paper.
- Viscosity—using Bohlin Instruments CVO 100 rheometer (Cirencester, UK), cone-plate 60 mm diameter and 1° measuring device, shear rate 600 s^{-1}.
- Coagulum content—after filtration of dispersion on 190 mesh net the solids remaining on the net were dried and weighed. Coagulum content (wt.%) was calculated from equation $m_c/m_d \times 100\%$ where m_c was mass of dry coagulum remaining on the net and m_d was mass of dispersion.
- Acrylic and styrene monomers, ethanol and D4 content—by GC (HP 5890 series II apparatus FID detector, Hewlett Packard, Palo Alto, CA, USA)

- Mechanical stability—lack or occurrence of separation during rotation in Hettich Universal 32R centrifuge (Westphalian, The Netherlands) at 4000 r.p.m. for 90 min was considered as good stability.
- Average particle size (nm), particle size distribution and zeta potential (mV)—light-scattering method using Malvern Zeta Sizer apparatus.
- Dispersion particles appearance—transmission electron microscope (TEM) Hitachi 2700 (Tokyo, Japan), dispersions were diluted 1000× with water (1 part of dispersion per 1000 parts of water) for taking pictures. High Angle Annular Dark Field (HAADF) mode also called "Z-contrast" was applied for processing the images reproduced in this paper.
- Minimum film-forming temperature (MFFT)—according to ISO 2115 [34] using Coesfeld apparatus equipped with temperature gradient plate. Temperature range: –3–50 °C.
- Glass transition temperature (T_g) of dispersion solids—by differential scanning calorimetry (DSC) (TA Instruments Q2000 apparatus, New Castle, DE, USA), heat–cool–heat regime, 20 °C/min.

2.4. Characterization of Coatings

Coatings were produced from dispersions by applying them on glass (for testing contact angle, hardness, adhesion or water resistance), aluminium plates (for testing elasticity) or on steel plates (for testing impact resistance and cupping) using 120 μm applicator. Drying was carried out for 30 min at 50 °C and then the coatings were seasoned in a climatic chamber at 23 °C and 55% relative humidity (R.H.) for 72 h. Since no continuous coating could be obtained in this procedure for SIL-ACR 1-B and SIL-ACR 2-B, the relevant dispersions were dried at 8 °C for 2 h and then seasoned as above. The resulting coatings were characterized by:

- Contact angle (water)—according to EN 828:2000, using KRUSS DSA 100E apparatus (KRÜSS GmbH, Hamburg, Germany). The measurements were conducted five times and the mean value was taken.
- Pendulum hardness (Koenig)—according to EN ISO 1522 [35]. The measurements were conducted seven times and the mean value was taken.
- Adhesion—according to EN ISO 2409 [36], the tests were repeated at least three times.
- Elasticity—according to EN ISO 1519 [37], the tests were repeated at least two times.
- Impact resistance (direct and reverse)—according to EN ISO 6272-1 [38], using Erichsen Variable Impact Tester Model 304 (Erichsen, Hemer, Germany). The measurements were conducted at least twice.
- Cupping—according to EN ISO 1520 [39], using Erichsen Cupping Tester (ERICHSEN GmbH & Co. KG, Hemer, Germany). The tests were repeated three times and the mean value was taken.
- Water resistance—glass Petri dishes of 50 mm diameter were filled with distilled water and placed upside-down on the coating, so the coating was covered with 7 mm thick layer of water. Assembles prepared this way were left for 72 h and appearance of coatings was examined for the bubbles size (S0—no bubbles, S2–S5—small to large size of bubbles) and density (0—no bubbles, 2–5 low to high density of bubbles) according to EN ISO 4628-2 [40]. Observation of changes of coating appearance after 6 days under water were also examined.
- Water vapour permeability—according to ASTM F1249 [41]. TotalPerm 063 (Extra Solution) apparatus was used. Tests were conducted at 23 °C for 0.35 mm thick film. Fomblin perfluorinated grease from Solvay Solexis (Brussels, Belgium) was applied to seal the test vessels. The measurements were repeated at least twice.
- Moreover, coatings applied on PET film were examined for surface structure by X-ray photoelectron spectroscopy (XPS)—ULVAC/PHYSICAL ELECTRONICS PHI5000 VersaProbe apparatus (Physical Electronics, Inc., Chigasaki, Japan).

2.5. Characterization of Films

- Percentage swell, i.e., change of the mass caused by soaking in water or organic solvent—ca. 0.12 g samples of film were weighed and placed in 40 mL H_2O or 40 mL toluene contained in closed glass cups and left for 20 h at 23 °C. Then the samples were taken out, delicately dried with filter paper and weighed. Percentage swell was calculated from the equation: % swell = m_1-m_0/m_0 × 100%, where m_0 = mass of the sample before test and m_1 = mass of the sample after test. The tests were repeated three times.
- Mechanical properties (tensile strength and elongation at break)—using Instron 3345 testing machine (Instron, Norwood, MA, USA) according to EN-ISO 527-1 [42] at a pulling rate of 50 mm/min on dumbbell-shaped specimens. The measurements were conducted five times and the mean value was used taken.

3. Results and Discussion

3.1. Properties of Dispersions

Properties of hybrid silicone-acrylic (SIL-ACR) and acrylic-silicone (ACR-SIL) dispersions prepared with SIL/ACR w/w 1/3 ratio, starting silicone resin dispersions (SIL 1 and SIL 2) and starting acrylic/styrene copolymer dispersions (ACR A and ACR B), are presented in Table 1. All hybrid dispersions were mechanically stable, slightly opalescent white liquids with low viscosity, pH in the range 5.8–6.3 and solids contents close to 42%. Coagulum content was at a very low level –0.04%–0.38%. Blends of starting SIL and ACR dispersions at w/w 1/3 ratio were also made, but the resulting dispersions were not mechanically stable and did not produced continuous coatings at room temperature.

3.1.1. Particle Size and Particle Size Distribution

For hybrid dispersions, particle size distribution was monomodal and rather narrow, although in most cases wider than that for starting SIL and ACR dispersions. Zeta potentials were all very low (i.e., very negative) which indicated good dispersion stability that was confirmed by mechanical stability tests.

The average particle size was distinctly higher for hybrid dispersions SIL-ACR than for starting SIL dispersion and almost the same for ACR-SIL than for starting ACR dispersion (see Figure 4) what could indicate formation of shell on starting SIL dispersion core particles during polymerization of ACR monomers and lack of formation of core-shell particle structure in the case of polymerization of SIL monomers in starting ACR dispersion. The comparison of particle size distribution patterns confirmed that assumption for ACR-SIL dispersions—see Figure 5a. As it can be seen in Figure 5b, in synthesis of SIL-ACR dispersions acrylic/styrene copolymer particles smaller than particles of starting SIL dispersion were probably formed along with core-shell SIL-ACR particles.

In general, average particle size was significantly higher for SIL-ACR dispersions than for ACR-SIL dispersions of the same composition of SIL and ACR parts—see Figure 6 where the particle size distribution of one of the SIL-ACR dispersions (SIL-ACR 2-B) and of the corresponding ACR-SIL dispersion (ACR-SIL B-2) is shown. The reason for that was higher particle size of starting SIL dispersions than for starting ACR dispersions.

Table 1. Properties of hybrid silicone-acrylic (SIL-ACR) and ACR-SIL dispersions and of starting SIL and ACR dispersions. In the case of starting ACR dispersions all properties were determined for dispersions neutralized after polymerization while ACR dispersions before neutralization (with pH ca. 3.0) were used in synthesis of ACR-SIL hybrids. N.A. = Not Applicable because dispersion did not produce continuous film at room temperature (R.T.).

Designation of Dispersions	pH	Solids Content %	Coagulum Content %	Viscosity at 23 °C mPa·s	Average Particle Size nm	Particle Size Distribution nm	Polydispersity	Zeta Potential mV	Minimum Film-Forming Temperature (MFFT) °C	T_g (Disp. Solids) °C
SIL 1	6.3	19.3	0.00	2	121.7	104–157	0.121	−50.9	N.A.	−117.33
SIL 2	6.2	18.5	0.00	3	128.6	116–154	0.086	−49.5	N.A.	−119.46
ACR A	6.2	50.1	<0.1	94	105.7	74–119	0.112	−51.0	+11.4	+17.17
ACR B	6.2	50.5	<0.1	98	112.2	108–135	0.066	−56.0	+32.7	+32.36
SIL-ACR 1-A	5.8	42.2	0.06	20	143.8	69–160	0.074	−57.7	+7.3	−126.66 / +17.66
ACR-SIL A-1	6.3	45.2	0.38	140	111.7	71–131	0.096	−55.2	−0.5	−130.54 / +14.58
SIL-ACR 1-B	6.3	42.3	0.11	24	140.3	69–190	0.072	−59.5	26.2	−129.09 / +30.87
ACR-SIL B-1	6.3	42.0	0.21	61	114.4	98–133	0.071	−55.0	16.0	−126.33 / +27.98
SIL-ACR 2-A	6.1	41.7	0.04	20	151.2	107–214	0.064	−59.2	−1.0 [1]	−132.94 / +17.29
ACR-SIL A-2	6.3	41.6	0.13	55	109.3	98–122	0.085	−47.2	−0.4	+16.13
SIL-ACR 2-B	6.3	41.4	0.05	16	149.9	125–156	0.057	−55.7	+26.0 [1]	−131.46 / +32.56
ACR-SIL B-2	6.3	42.2	0.24	56	115.9	104–130	0.069	−53.6	+14.8	+20.04

[1] For these dispersions also maximum film forming temperature was observed. It was +10.5 °C for SIL-ACR 2-A and +36.2 °C for SIL-ACR 2-B.

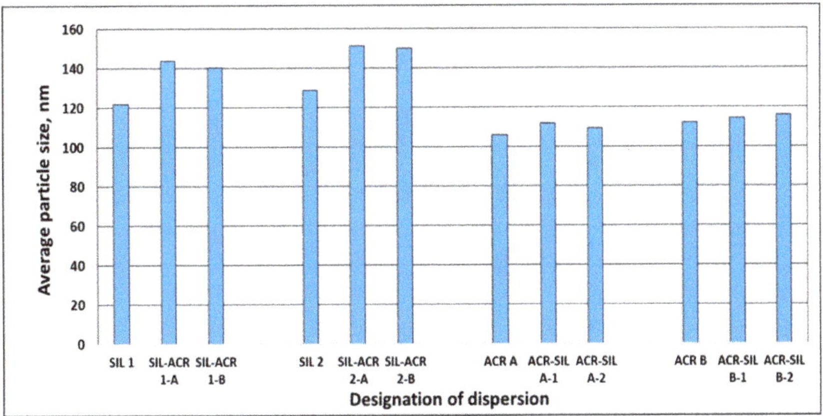

Figure 4. Comparison of average particle size of SIL and ACR dispersions and hybrid SIL-ACR and ACR-SIL dispersions.

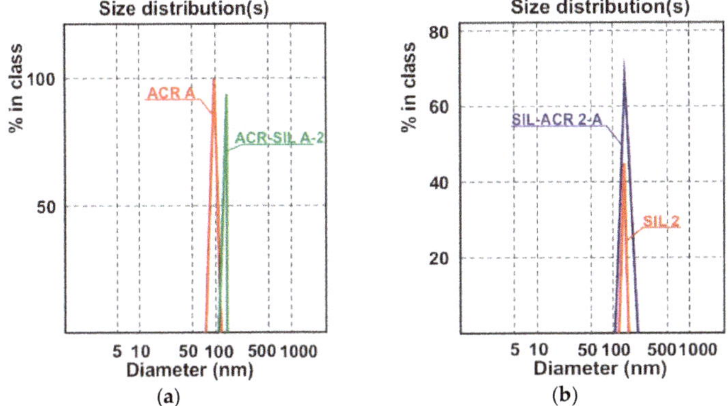

Figure 5. Comparison of particle size distribution patterns of hybrid ACR-SIL A-2 dispersion and starting ACR A dispersion (**a**) and of hybrid SIL-ACR 2-A dispersion and starting SIL 2 dispersion (**b**). X-axis is logarithmic.

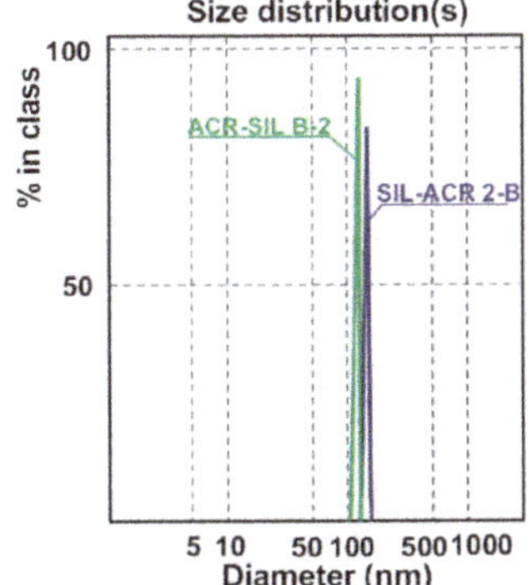

Figure 6. Comparison of particle size distribution patterns for SIL-ACR and ACR-SIL dispersions of the same composition of SIL and ACR parts. X-axis is logarithmic.

3.1.2. Particle Structure

In Figure 7 the structure of hybrid dispersion particles of SIL-ACR and ACR-SIL dispersions determined by TEM is shown. As can be seen in Figure 7a, in the case of SIL-ACR dispersion coalescence of particles proceeded during testing, so the TEM image shows a tiny piece of film rather than the single particle, but it is clear that well defined silicone resin particles (lighter shade) are surrounded by acrylic/styrene copolymer phase (darker shade). Individual particles can be identified better in Figure 7b where lower magnification was used and it can be concluded that a kind of "fruit cake" particle structure where a few "cores" made of one polymer are surrounded by continuous mass of the other polymer was formed during polymerization of ACR monomers in SIL dispersion. In the case of ACR-SIL, dispersion coalescence of particles during testing also proceeded. While both individual particles and aggregates of silicone resin particles and acrylic/styrene copolymer particles were present, it was also possible to identify in TEM images abundant single particles of specific structure shown in Figure 7c. In this structure kinds of spheres made of silicone resin (lighter shade) were embedded in the mass of acrylic/styrene copolymer (darker shade). It can be anticipated that in the course of synthesis of ACR-SIL hybrid dispersions silicone monomers penetrated into acrylic/styrene copolymer particles and after completion of polymerization a kind of sphere of silicone resin was formed because of lack of compatibility of acrylic/styrene copolymer and silicone resin. Such a particle structure called an "embedded sphere" has been found also earlier in polyurethane-acrylic/styrene hybrid dispersions [4].

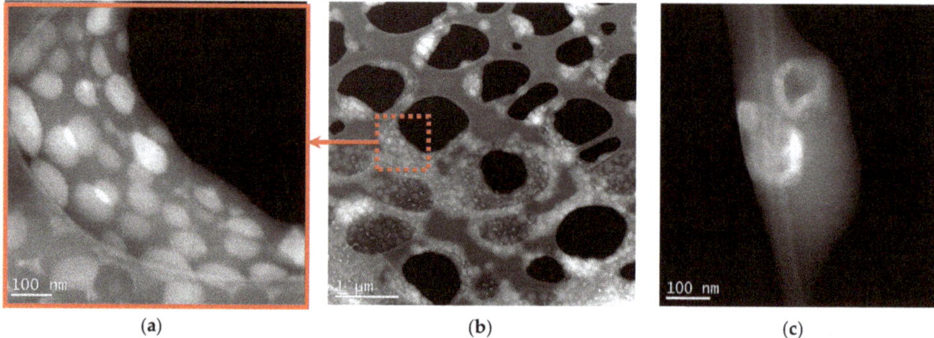

(a)	(b)	(c)

Figure 7. Structure of hybrid dispersion particles settled on the micromesh net as determined by transmission electron microscopy (TEM): (**a**) SIL-ACR dispersion, higher magnification, (**b**) SIL-ACR dispersion, lower magnification, (**c**) ACR-SIL dispersion, higher magnification. Lighter shade represents silicone resin and darker shade–acrylic/styrene copolymer.

Lack of formation of core-shell ACR-SIL hybrid particles in the course of polymerization of silicone monomers in acrylic/styrene copolymer dispersion could have been expected since it was clear from the review of available literature on that subject [11] that only if special approaches were applied to synthesis (e.g., functionalization of acrylic particle surface with silane and hydrolysis of alkoxysilane groups prior to polymerization [22]) the particles with acrylic polymer core and silicone shell could be obtained.

3.1.3. Minimum Film-Forming Temperature (MFFT)

As it can be seen in Figure 8 MFFT values determined for ACR-SIL hybrid dispersions were much lower than for starting ACR dispersion and lower than for SIL-ACR dispersions of the same SIL and ACR parts composition what can be explained by the fact that only a fraction of particles of ACR-SIL dispersion hybrid structure exhibited a hybrid morphology shown in Figure 6b and the presence of separate silicone resin particles resulted in lower MFFT.

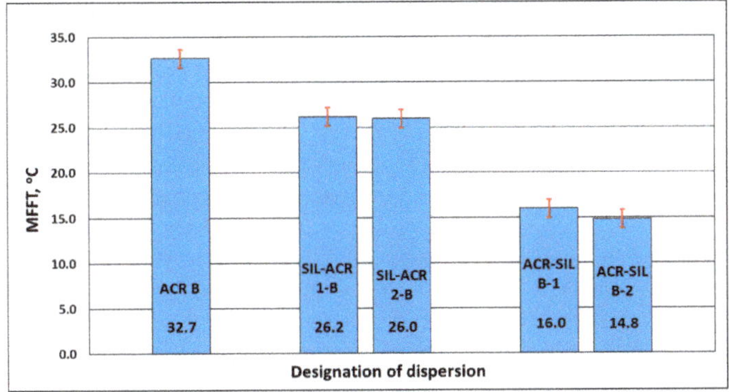

Figure 8. Comparison of MFFT determined for hybrid dispersions SIL-ACR and ACR-SIL. MFFT of starting ACR dispersion used for synthesis of ACR-SIL dispersions is also shown.

3.1.4. Glass Transition Temperature (T_g)

DSC results showed that hybrid dispersion solids usually exhibited two T_{gs}: one corresponding to SIL part at c.a. $-120\ °C$ and the other corresponding to ACR part in the range of ca. $15–30\ °C$, depending on the T_g of starting acrylic/styrene copolymer—see Figure 9 where DSC patterns determined for starting SIL and ACR dispersions and for SIL-ACR and ACR-SIL dispersions having the same composition of ACR and SIL parts are presented.

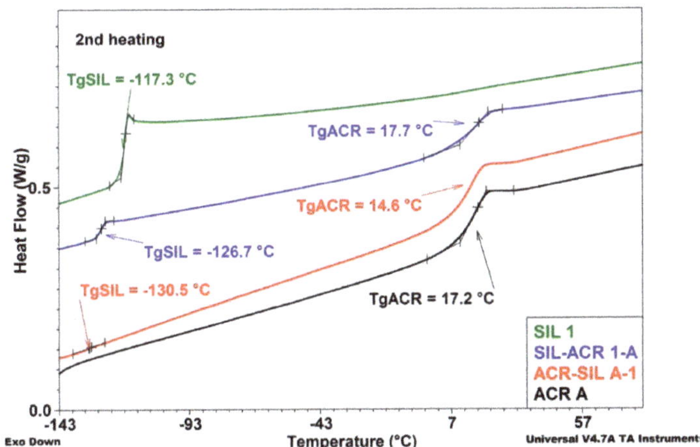

Figure 9. Differential scanning calorimetry (DSC) patterns determined for starting SIL and ACR dispersions and for SIL-ACR and ACR-SIL dispersions having the same composition of ACR and SIL parts.

Only for two dispersions (ACR-SIL A-2 and ACR-SIL B-2) just one T_g was detected at around 16 and 20 °C, respectively, what suggested that in the case of these two dispersions the particle structure was rather uniform and no separate silicone resin particles were formed. This phenomenon can be explained by the fact that in these two dispersions silicone monomers (D4 + ethoxy-functional silane) mixture that was polymerized in acrylic/styrene copolymer dispersion contained much more VTES (more polar) and did not contain MTES (less polar), so penetration into acrylic/styrene copolymer particles was easier and grafting of VTES on acrylic/styrene copolymer and formation of "embedded sphere" structures shown in Figure 7c were much more probable.

It was also interesting that T_{gs} of SIL and ACR parts of all hybrid dispersion solids where two glass transitions were detected were significantly lower than T_{gs} of starting SIL and ACR dispersions solids. Decrease in T_g of ACR part can be explained by plasticizing effect of modification with silicone resin. However, in order to clarify why decrease in T_g of SIL partly occurred, more insight is needed to the processes which took place in the course of both silicone monomers polymerization in acrylic/styrene copolymer dispersion and acrylic/styrene monomers polymerization in silicone resin dispersion. The key assumption (confirmed by the hybrid particle structures) is that in hybrid dispersion particles silicone resin particles are "trapped" within a mass of acrylic/styrene copolymer, so D4 and higher oligodimethylsiloxane cycles (e.g., D5) which are always present in SIL dispersions [43] and are also formed in synthesis of hybrid ACR-SIL dispersions are also "trapped" and therefore cannot be released during drying and may plasticize the silicone resin contained in dispersion solids. That "trapping" of silicone resin in acrylic/styrene copolymer part of hybrid dispersion particles should be more distinct if the SIL part contained more VTES because of possibility of grafting the decrease in T_g should be more distinct for hybrid dispersions ACR-SIL 1-A and ACR-SIL 1-B than for hybrid dispersions ACR-SIL 2-A and ACR-SIL 2-B. Comparison of the relevant T_g values in Table 1 confirmed that this was actually the case.

3.2. Properties of Coatings and Films

Properties of coatings and films obtained from hybrid silicone-acrylic (SIL-ACR) and acrylic-silicone (ACR-SIL) dispersions prepared with SIL/ACR w/w 1/3 ratio, starting silicone resin dispersions (SIL 1 and SIL 2) and starting acrylic/styrene copolymer dispersions (ACR A and ACR B) are presented in Table 2. Some hybrid dispersions and starting silicone resin dispersions did not form mechanically strong continuous coatings or films, but certain properties like e.g., contact angle or % swell could be determined by casting layers which, after drying, formed mechanically weak coatings or films.

It is essential that for all hybrid dispersions the key coating properties that were expected to improve as compared to acrylic/styrene copolymer dispersions (contact angle, water vapor permeability and water resistance) actually did improve significantly. Mechanical properties of coatings (e.g., impact resistance or elasticity) also improved, but hardness decreased what could be expected. The same trend was reflected in film properties—increase in elongation at break was accompanied by a decrease in tensile strength.

3.2.1. Surface Properties

The high contact angle of coatings is important since it means high surface hydrophobicity and, consequently, lower water uptake and lower dirt deposition [5]. As can be seen in Table 2, all coatings obtained from hybrid SIL-ACR and ACR-SIL dispersions showed high contact angles in the range of 80–90° while contact angles recorded for films obtained from starting ACR dispersions were quite low (ca. 30°). It is worth to note that contact angles recorded for coatings produced from ACR-SIL dispersions were generally higher than those recorded for coatings produced from SIL-ACR dispersions (see Figure 10) what indicates that in the former case more silicone migrated to the coating surface.

Migration of silicone to the coating surface observed for coatings containing silicones was described in the earlier papers, e.g., [32,44,45] and was fully confirmed by XPS also for coatings obtained from SIL-ACR and ACR-SIL hybrid dispersions. In Figure 11 the percentage of Si in the layers close to coating surface as determined by XPS for hybrid SIL-ACR and ACR-SIL dispersions is plotted against distance from the surface. It is clear from Figure 11 that in the coatings obtained from hybrid dispersions silicone migrated to coating surface and that migration was different for coatings obtained from ACR-SIL dispersions than for those obtained from SIL-ACR dispersions, most probably due to "trapping" of silicone resin in acrylic/styrene copolymer particles in the latter coating.

Table 2. Properties of coatings and films made from hybrid SIL-ACR and ACR/SIL dispersions and of starting SIL and ACR dispersions. N.A. = Not Applicable because dispersion did not produce continuous coating or/and film at R.T. Although SIL-1 and SIL-2 dispersions did not produce continuous a mechanically strong coatings, it was possible to measure their contact angles on very mechanically weak coats that were cast on glass Petri dishes.

Designation of Dispersions	Contact Angle (H_2O) (°)	Water Vapour Permeability $g/m^2/24h$	Water Resistance after 72 h	Impact Resistance (direct) J	Impact Resistance (reverse) J	Cupping mm	Elasticity (Rod Diameter 2 mm)	Hardness (Koenig)	Adhesion to Glass	Swell in H_2O %	Swell in Toluene %	Tensile Strength MPa	Elongation at Break %
SIL 1	111				N.A.					18	202	N.A.	
SIL 2	104				N.A.					10	387	N.A.	
ACR A	30	28.1	>5(S5) 5(S2)	2.0	19.6	10.7	passed	0.082	5	14	1156	4.2	1000
ACR B	35	15.6	Medium whitening 5(S2)	2.0	0	10.7	failed	0.458	5	15	1547	12.0	340
SIL-ACR 1-A	83	56.5	Medium whitening 0(S0)	9.8	19.6	11.6	passed	0.040	3	26	591	2.1	773
ACR-SIL A-1	95	45	Light whitening 5(S2)	15.7	19.6	11.0	passed	0.022	2	21	1050	0.8	1851
SIL-ACR 1-B	81	64.5	Light whitening 0(S0)	0	0	10.9	passed	0.085	5	11	561	4.3	11
ACR-SIL B-1	92	34.6	Medium whitening	5.9	19.6	10.9	passed	0.050	5	26	905	3.1	1015
SIL-ACR 2-A	77				N.A.					20	605		
ACR-SIL A-2	92	39.4	0(S0) Medium whitening	13.7	19.6	10.5	passed	0.034	5	16	1112	0.9	1516
SIL-ACR 2-B	85				N.A.					9	665	N.A.	
ACR-SIL B-2	82	28.0	0(S0) Medium whitening	3.9	19.6	11.0	passed	0.058	5	31	991	3.2	947

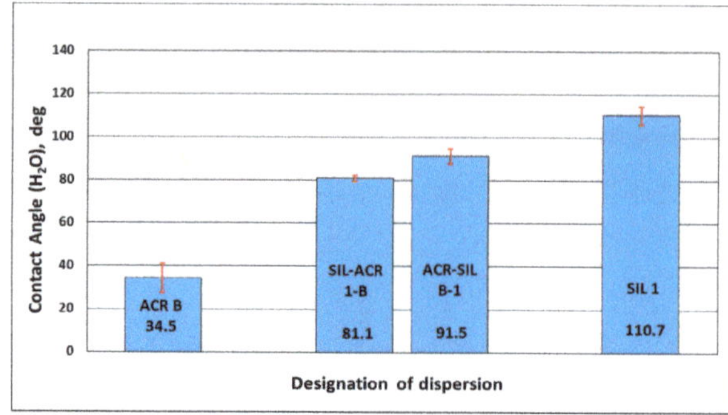

Figure 10. Comparison of contact angle values determined for coatings obtained from starting SIL and ACR dispersions and corresponding hybrid SIL-ACR and ACR-SIL dispersions having the same composition of ACR and SIL parts. Contact angle determined for starting acrylic/styrene copolymer dispersion (ACR B) is also shown.

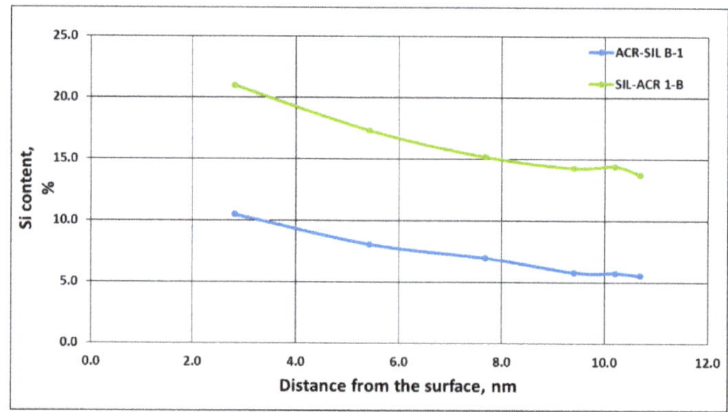

Figure 11. Decrease in Si content with distance from coating surface determined by XPS for coatings obtained from SIL-ACR and ACR-SIL dispersions.

3.2.2. Water Resistance

Good water resistance of architectural paints is crucial since it ensures longer life of the paint and better comfort of the building walls (lack of water uptake) if combined with high water vapour permeability. Therefore, determination of the water resistance of coatings produced from dispersions which are intended to be applied as binders for architectural paints seems to be very important test. In our investigations we measured water resistance of coatings obtained from starting ACR dispersions and from SIL-ACR and ACR-SIL dispersions using our own method partly described in Section 2.4 and the results were assessed based on EN ISO 4628-2 [40]. All coatings made from hybrid dispersions exhibited better water resistance than those produced from starting ACR dispersions and it was significantly better for coatings obtained from ACR-SIL dispersions than from SIL-ACR dispersions—see Figure 12 where photos of coatings produced from different dispersions and left under water for 6 days are shown.

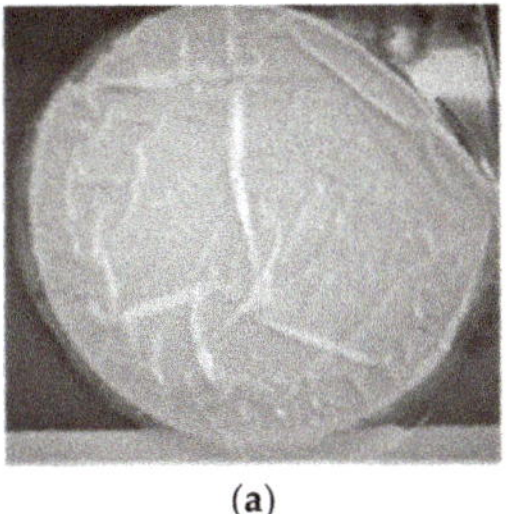 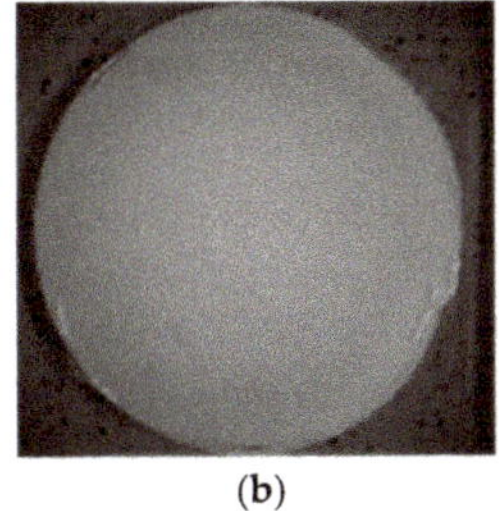 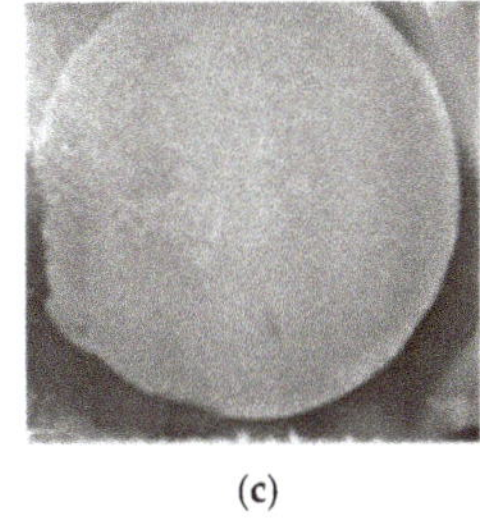

(a)   (b)   (c)

Figure 12. Comparison of water resistance of starting acrylic/styrene copolymer dispersion. ACR A (**a**), acrylic-silicone dispersion (ACR-SIL A-1) (**b**) and corresponding silicone-acrylic dispersion SIL-ACR (**c**). Samples were kept under water for 6 days. ACR A—deterioration of coating occurred, ACR-SIL A-1—coating did not change except for light whitening, SIL-ACR 1-A—coating changed significantly–numerous small bubbles.

3.2.3. Swell in Water and in Toluene

As can be concluded from Table 2 percent of swell in water was very similar for all films (despite of differences in water resistance of coatings) and was quite low (ca. 20%) while swell in toluene that can be considered as a measure of crosslinking density (higher swell means lower crosslinking density) was much higher for films made from ACR dispersions than for films made from SIL dispersions, and also much higher for films made from hybrid ACR-SIL dispersions than for films made from SIL-ACR dispersions—see the relevant comparison in Figure 13.

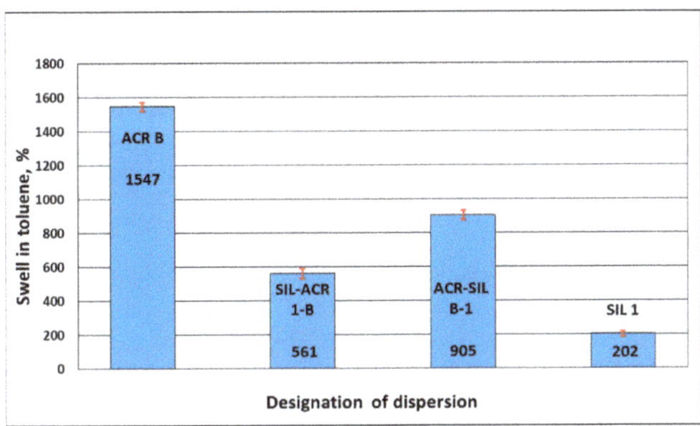

Figure 13. Comparison of % swell in toluene determined for starting silicone resin dispersion (SIL-1), starting acrylic/styrene copolymer dispersion (ACR B) and hybrid dispersions ACR-SIL B-1 and SIL-ACR 1-B having the same composition of SIL and ACR parts.

The difference between crosslinking density of films (i.e., also for coatings) made from ACR-SIL and SIL-ACR dispersions having the same composition of ACR and SIL parts can be explained by a higher possibility of grafting of acrylic/styrene monomers on silicone resin than of grafting VTES on acrylic/styrene copolymer. Another reason can be a higher possibility of trapping of partly crosslinked silicone resin inside particles made of acrylic/styrene copolymer in the case of films made from SIL-ACR dispersions than in the case of films made from ACR-SIL dispersions—see the discussion of hybrid dispersions particle structures contained in Section 3.1.2.

3.2.4. Water Vapour Permeability

As has already been pointed out in Section 3.2.2, good architectural paint should exhibit not only good water resistance, but also good water vapour permeability. This positive combination of properties can be achieved in practice only for paints based on silicone-acrylic binders because silicone polymers are characterized by good permeability of gases due to high mobility of poly(dimethylsiloxane) chains. It was proved in our study that coatings produced from hybrid SIL-ACR and ACR-SIL dispersions showed higher water vapour permeability than those produced from starting ACR dispersions—see Figure 14.

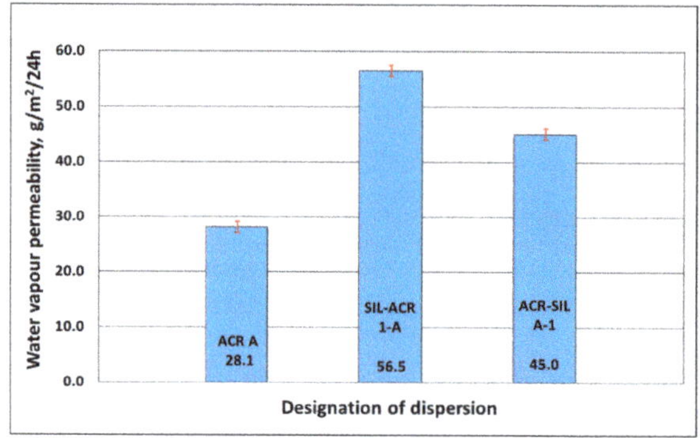

Figure 14. Comparison of water vapour permeability determined for starting acrylic/styrene copolymer dispersion (ACR A) and hybrid dispersions ACR-SIL A-1 and SIL-ACR 1-A having the same composition of SIL and ACR parts.

It can be noted from the results presented in Figure 14 that water vapour permeability was better for coatings obtained from SIL-ACR dispersions than from ACR-SIL dispersions, probably because of differences in coating structure that resulted from differences in dispersion particle structure.

3.2.5. Mechanical Properties

If the results of testing the mechanical properties of coatings and films produced from hybrid ACR-SIL and SIL-ACR dispersions presented in Table 2 are compared with mechanical properties of coatings produced from starting ACR dispersions, it is clear that modification with silicone led generally to less brittle coatings, especially in the case of starting dispersion ACR A. The most spectacular difference was in the (direct) impact resistance of coatings—see Figure 15.

For coatings and films produced from starting dispersion ACR B and hybrid coatings and films where ACR B composition of monomers was applied in synthesis of the relevant dispersions, the results of mechanical tests were much less convincing, presumably because T_g of ACR B was quite high (over 30 °C). Cupping test results were good for all coatings and in direct elasticity measurements, only coatings produced from starting dispersion ACR B failed. Elongation at break increased for some films made from hybrid dispersions as compared to films made from starting ACR dispersions and decreased for some others (specifically for these produced from hybrid dispersions with particles having ACR B composition of ACR part) and tensile strength decreased for all films where this could be expected taking into account plasticizing effect of silicone resin. Much higher elongation at break and much lower tensile strength observed for films made from ACR-SIL dispersions than from SIL-ACR dispersions can be explained by a different supramolecular structure of films that results from different morphology of hybrid dispersion particles (see Figure 7) that coalesce to produce these films in the process of air-drying of dispersions.

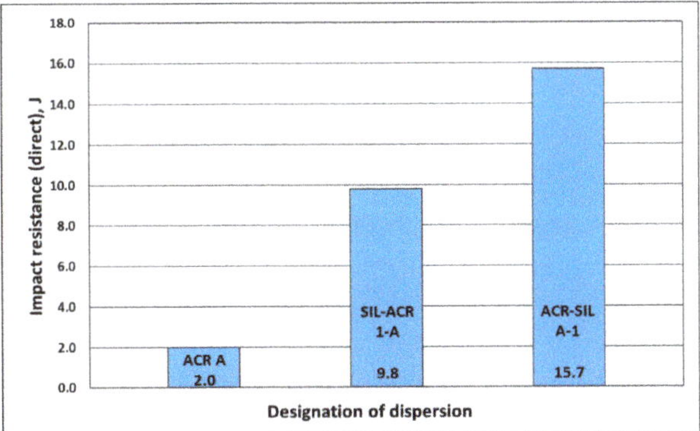

Figure 15. Comparison of impact resistance (direct) determined for coatings obtained from starting acrylic/styrene copolymer dispersion (ACR A) and hybrid dispersions ACR-SIL A-1 and SIL-ACR 1-A having the same composition of SIL and ACR parts.

4. Conclusions

Simultaneous synthesis of aqueous silicone-acrylic and acrylic-silicone hybrid dispersions (SIL-ACR and ACR-SIL) by (1) emulsion polymerization of acrylic/styrene monomers (BA, ST, KA and MA) mixtures of different composition (ACR A and ACR B) in aqueous dispersions of silicone resins of different composition (SIL 1 and SIL 2) and (2) emulsion polymerization of silicone monomers (D4, VTES and MTES) mixtures of different composition (SIL 1 and SIL 2) in aqueous dispersions of acrylic/styrene copolymers (ACR A and ACR B) was successfully conducted. Hybrid dispersions had good mechanical stability, low minimum film-forming temperature and particle size in the range of 100–150 nm, narrow particle size distribution, and contained very little of coagulate. TEM investigation of hybrid dispersions particle structure revealed that particles of SIL-ACR dispersions exhibited "fruit cake" structure while particles of ACR-SIL dispersions showed "embedded sphere" structure. For most of the dispersions two separate T_{gs} of dispersion solids (one for SIL part and the other for ACR part) that were detected by DSC were lower than T_{gs} of corresponding starting SIL and ACR dispersions while single T_g was detected for two of them. These differences were explained by differences in dispersion particle structure.

Most of the hybrid dispersions formed mechanically strong continuous coatings and films. As compared to coatings obtained from starting ACR dispersions, those obtained from hybrid dispersions showed much higher contact angles, much better water resistance and water vapour permeability and exhibited much better impact resistance. Different coating properties were observed when coatings were produced from SIL-ACR and ACR-SIL dispersions having the same composition of ACR and SIL parts, which most probably resulted from different structure of dispersions particles. Films produced from hybrid dispersions were less brittle than those produced from starting ACR dispersions. Determinations of % swell in toluene measured for films produced from hybrid dispersions revealed the difference between crosslinking density of films (i.e., also for coatings) made from ACR-SIL and SIL-ACR dispersions having the same composition of ACR and SIL parts, which was explained by higher possibility of grafting of acrylic/styrene monomers on silicone resin than of grafting VTES on acrylic/styrene copolymer. The authors believe that the selected hybrid dispersions described in this paper can be applied as binders in the formulation of architectural paints that will be characterized by high water resistance and high surface hydrophobicity combined with high water vapour permeability.

Author Contributions: Conceptualization, J.K. and J.T.; Methodology, J.T., W.D., I.O.-K., A.K., K.S. and J.P.; Investigation, J.T., W.D. and A.K.; Writing-Original Draft Preparation, J.K.; Writing-Review & Editing, J.T. and W.D.; Visualization, J.P., W.D. and J.T.; Supervision, J.K., D.W. and M.W.; Project Administration and Funding Aquisition, J.K.

Funding: This research was funded by Polish State R&D Centre (NCBiR, No. PBS/B1/8/2015).

Acknowledgments: The authors wish to thank Piotr Bazarnik from Warsaw University for conducting TEM studies, and Janusz Sobczak from Polish Academy of Sciences for conducting XPS studies. The assistance of Colleagues from the Industrial Chemistry Research Institute in testing mechanical properties of films, water vapour permeability and dispersion particle size distribution, is also acknowledged.

Conflicts of Interest: The authors declare no conflict of interest.

References

1. Rodríguez, R.; de Las Heras Alarcón, C.; Ekanayake, P.; McDonald, P.J.; Keddie, J.L.; Barandiaran, M.J.; Asua, J.M. Correlation of silicone incorporation into hybrid acrylic coatings with the resulting hydrophobic and thermal properties. *Macromolecules* **2008**, *41*, 8537–8546. [CrossRef]
2. Kickelbick, G. Introduction to hybrid materials. In *Hybrid Materials: Synthesis, Characterization and Applications*, 1st ed.; Wiley-VCH Verlag GmbH & Co. KGaA: Weinheim, Germany, 2007; pp. 1–46.
3. Castelvetro, V.; de Vita, C. Nanostructured hybrid materials from aqueous polymer dispersions. *Adv. Colloid Interface Sci.* **2004**, *108–109*, 167–185. [CrossRef]
4. Kozakiewicz, J.; Koncka-Foland, A.; Skarżyński, J.; Legocka, I. Synthesis and characterization of aqueous hybrid polyurethane-urea-acrylic/styrene polymer dispersions. In *Advances in Urethane Science and Technology*, 1st ed.; Klempner, D., Frisch, K.C., Eds.; RAPRA Technology Ltd.: Shrewsbury, UK, 2001; pp. 261–334.
5. Khanjani, J.; Pazokifard, S.; Zohuriaan-Mehr, M.J. Improving dirt pickup resistance in waterborne coatings using latex blends of acrylic/PDMS polymers. *Prog. Org. Coat.* **2017**, *102*, 151–166. [CrossRef]
6. Peruzzo, P.J.; Anbinder, P.S.; Pardini, O.R.; Vega, J.; Costa, C.A.; Galembeck, F.; Amalvy, J.I. Waterborne polyurethane/acrylate: Comparison of hybrid and blend systems. *Prog. Org. Coat.* **2011**, *72*, 429–437. [CrossRef]
7. Ma, J.Z.; Liu, Y.H.; Bao, Y.; Liu, J.L.; Zhang, J. Research advances in polymer emulsion based on "core-shell" structure design. *Adv. Colloid Interface Sci.* **2013**, *197–198*, 118–131. [CrossRef]
8. Mittal, V. *Advanced Polymer Nanoparticles: Synthesis and Surface Modification*, 1st ed.; CRC Press: Boca Raton, FL, USA, 2010.
9. Guyot, A.; Landfester, K.; Schork, F.J.; Wang, C. Hybrid polymer latexes. *Prog. Polym. Sci.* **2007**, *32*, 1439–1461. [CrossRef]
10. Ghosh Chaudhuri, R.; Paria, S. Core/shell nanoparticles: Classes, properties, synthesis mechanisms, characterization, and applications. *Chem. Rev.* **2012**, *112*, 2373–2433. [CrossRef]
11. Kozakiewicz, J.; Ofat, I.; Trzaskowska, J. Silicone-containing aqueous polymer dispersions with hybrid particle structure. *Adv. Colloid Interface Sci.* **2015**, *223*, 1–39. [CrossRef]
12. Holmes, D. Controlling the morphology of composite latex particles. *Inquiry J.* **2005**, *4*. Available online: https://scholars.unh.edu/inquiry_2005/4/ (accessed on 25 December 2018).
13. Winzor, C.L.; Sundberg, D.C. Conversion dependent morphology predictions for composite emulsion polymers: 1. Synthetic lattices. *Polymer* **1992**, *33*, 3797–3809. [CrossRef]
14. Chen, Y.C.; Dimonie, V.; El-Aasser, M.S. Interfacial phenomena controlling particle morphology of composite latexes. *J. Appl. Polym. Sci.* **1991**, *42*, 1049–1063. [CrossRef]
15. Ferguson, C.J.; Russel, G.T.; Gilbert, R.G. Modelling secondary particle formation in emulsion polymerisation: Application to making core–shell morphologies. *Polymer* **2002**, *43*, 4557–4570. [CrossRef]
16. Sundberg, D.S.; Durant, Y.G. Latex particle morphology, fundamental aspects: A review. *Polym. React. Eng.* **2003**, *11*, 379–432. [CrossRef]
17. Eduok, U.; Faye, O.; Szpunar, J. Recent developments and applications of protective silicone coatings: A review of PDMS functional materials. *Prog. Org. Coat.* **2017**, *111*, 124–163. [CrossRef]
18. He, W.D.; Cao, C.T.; Pan, C.Y. Formation mechanism of silicone rubber particles with core-shell structure by seeded emulsion polymerization. *J. Appl. Polym. Sci.* **1996**, *61*, 383–388. [CrossRef]

19. He, W.D.; Pan, C.Y. Influence of reaction between second monomer and vinyl group of seed polysiloxane on seeded emulsion polymerization. *J. Appl. Polym. Sci.* **2001**, *80*, 2752–2758. [CrossRef]

20. Lin, M.; Chu, F.; Gujot, A.; Putaux, J.L.; Bourgeat-Lami, E. Silicone-polyacrylate composite latex particles. Particles formation and film properties. *Polymer* **2005**, *46*, 1331–1337. [CrossRef]

21. Shen, J.; Hu, Y.; Li, L.X.; Sun, J.W.; Kan, C.Y. Fabrication and characterization of polysiloxane/polyacrylate composite latexes with balanced water vapor permeability and mechanical properties: Effect of silane coupling agent. *J. Coat. Technol. Res.* **2018**, *15*, 165–173. [CrossRef]

22. Kan, C.Y.; Kong, X.Z.; Yuan, Q.; Liu, D.S. Morphological prediction and its application to the synthesis of polyacrylate/polysiloxane core/shell latex particles. *J. Appl. Polym. Sci.* **2001**, *80*, 2251–2258. [CrossRef]

23. Bourgeat-Lami, E.; Tissot, I.; Lefevbre, F. Synthesis and characterization of SiOH-functionalized polymer latexes using methacryloxy propyl trimethoxysilane in emulsion polymerization. *Macromolecules* **2002**, *35*, 6185–6191. [CrossRef]

24. Kozakiewicz, J.; Rościszewski, P.; Rokicki, G.; Kołdoński, G.; Skarżyński, J.; Koncka-Foland, A. Aqueous dispersions of siloxane-acrylic/styrene copolymers for use in coatings–preliminary investigations. *Surf. Coat. Int. Part B* **2001**, *84*, 301–307. [CrossRef]

25. Kan, C.Y.; Zhu, X.L.; Yuan, Q.; Kong, X.Z. Graft emulsion copolymerization of acrylates and siloxane. *Polym. Adv. Technol.* **1997**, *8*, 631–633. [CrossRef]

26. Li, W.; Shen, W.; Yao, W.; Tang, J.; Xu, J.; Jin, L.; Zhang, J.; Xu, Z. A novel acrylate-PDMS composite latex with controlled phase compatibility prepared by emulsion polymerization. *J. Coat. Technol. Res.* **2017**, *14*, 1259–1269. [CrossRef]

27. Xu, W.; An, Q.; Hao, L.; Zhang, D.; Zhang, M. Synthesis and characterization of self-crosslinking fluorinated polyacrylate soap-free lattices with core-shell structure. *Appl. Surf. Sci.* **2013**, *268*, 373–380. [CrossRef]

28. Hao, G.; Zhu, L.; Yang, W.; Chen, Y. Investigation on the film surface and bulk properties of fluorine and silicon contained polyacrylate. *Prog. Org. Coat.* **2015**, *85*, 8–14. [CrossRef]

29. Li, J.; Zhong, S.; Chen, Z.; Yan, X.; Li, W.; Yi, L. Fabrication and properties of polysilsesquioxane-based trilayer core-shell structure latex coatings with fluorinated polyacrylate and silica nanocomposite as the shell layer. *J. Coat. Technol. Res.* **2018**, *15*, 1077–1078. [CrossRef]

30. Xu, W.; Hao, L.; An, Q.; Wang, X. Synthesis of fluorinated polyacrylate/polysilsesquioxane composite soap-free emulsion with partial trilayer core-shell structure and its hydrophobicity. *J. Polym. Res.* **2015**, *22*, 20. [CrossRef]

31. Kozakiewicz, J.; Ofat, I.; Legocka, I.; Trzaskowska, J. Silicone-acrylic hybrid aqueous dispersions of core–shell particle structure and corresponding silicone-acrylic nanopowders designed for modification of powder coatings and plastics. Part I–Effect of silicone resin composition on properties of dispersions and corresponding nanopowders. *Prog. Org. Coat.* **2014**, *77*, 568–578. [CrossRef]

32. Kozakiewicz, J.; Ofat, I.; Trzaskowska, J.; Kuczynska, H. Silicone-acrylic hybrid aqueous dispersions of core–shell particle structure and corresponding silicone-acrylic nanopowders designed for modification of powder coatings and plastics. Part II: Effect of modification with silicone-acrylic nanopowders and of composition of silicone resin contained in those nanopowders on properties of epoxy-polyester and polyester powder coatings. *Prog. Org. Coat.* **2015**, *78*, 419–428. [CrossRef]

33. Pilch-Pitera, B.; Kozakiewicz, J.; Ofat, I.; Trzaskowska, J.; Spirkova, M. Silicone-acrylic hybrid aqueous dispersions of core–shell particle structure and corresponding silicone-acrylic nanopowders designed for modification of powder coatings and plastics. Part III: Effect of modification with selected silicone-acrylic nanopowders on properties of polyurethane powder coatings. *Prog. Org. Coat.* **2015**, *78*, 429–436. [CrossRef]

34. *ISO 2115 Polymer Dispersions–Determination of White Point Temperature and Minimum Film-Forming Temperature*; International Organization for Standardization: Geneva, Switzerland, 1996.

35. *ISO 1552 Paints and Varnishes–Pendulum Damping Test*; International Organization for Standardization: Geneva, Switzerland, 2006.

36. *ISO 2409 Paints and Varnishes–Cross-Cut Test*; International Organization for Standardization: Geneva, Switzerland, 2013.

37. *ISO 1519 Paints and Varnishes–Bend Test (Cylindrical Mandrel)*; International Organization for Standardization: Geneva, Switzerland, 2011.

38. *ISO 6272-1 Paints and Varnishes–Rapid-Deformation (Impact Resistance) Tests–Part 1: Falling-Weight Test, Large-Area Indenter*; International Organization for Standardization: Geneva, Switzerland, 2011.

39. *ISO 1520 Paints and Varnishes–Cupping Test*; International Organization for Standardization: Geneva, Switzerland, 2006.

40. *ISO 4628-2 Paints and Varnishes–Evaluation of Degradation of Coatings–Designation of Quantity and Size of Defects, and of Intensity of Uniform Changes in Appearance–Part 2: Assessment of Degree of Blistering*; International Organization for Standardization: Geneva, Switzerland, 2016.

41. *ASTM F1249 Standard Test Method for Water Vapor Transmission Rate Through Plastic Film and Sheeting Using a Modulated Infrared Sensor*; ASTM International: West Conshohocken, PA, USA, 2013.

42. *ISO 527-1 Plastics–Determination of Tensile Properties–Part 1: General Principles*; International Organization for Standardization: Geneva, Switzerland, 2012.

43. Liu, Y. *Silicone Dispersions*, 1st ed.; CRC Press: Boca Raton, FL, USA, 2016.

44. Mequanint, K.; Sanderson, R. Self-assembling of metal coatings from phosphate and siloxane-modified polyurethane dispersions: An analysis of the coating interface. *J. Appl. Polym. Sci.* **2003**, *88*, 893–899. [CrossRef]

45. Ofat, I.; Kozakiewicz, J. Modification of epoxy-polyester and polyester powder coatings with silicone-acrylic nanopowders–effect on surface properties of coatings. *Polimery* **2014**, *59*, 643–649. [CrossRef]

Article

Assessing of New Coatings for Iron Artifacts Conservation by Recurrence Plots Analysis

Paola Roncagliolo Barrera [1],*, Francisco Javier Rodríguez Gómez [1] and Esteban García Ochoa [2]

[1] Departamento de Ingeniería Metalúrgica, Facultad de Química, Universidad Nacional Autónoma de México, C.U., Ciudad de Mexico, C.P. 04510, Mexico; fxavier@unam.mx

[2] Centro de Investigación en Corrosión (CICORR), Universidad Autónoma de Campeche, Av. A. Melgar s/n, Col. Buenavista, Campeche, Cam, C.P. 24030, Mexico; estebangarci@gmail.com

* Correspondence: paolaroncagliolo@gmail.com; Tel.: +52-555-622-5225

Received: 16 November 2018; Accepted: 23 December 2018; Published: 26 December 2018

Abstract: Cast iron has stood for centuries of invention. It is a very versatile and durable material. Coating systems are a low-maintenance protection method. The purpose of this research is to increase the Paraloid coating's resistance when applied to iron in high humidity atmospheres, with the addition of caffeine (1,3,7-dimethylxanthine) and nicotine (S)-3-(1-methylpyrrolidin-2-yl) pyridine as corrosion inhibitors; the resistance of protection versus exposure time will be evaluated by using electrochemical noise. A statistical analysis of the electrochemical noise signals was carried out. Recurrence plots were used as a powerful tool in the analysis to complement the data obtained and they predicted the evaluation of coatings behaviors performance versus time. The outcomes show that the addition of inhibitors increases and improves the performance as a temporary protection of Paraloid and that protection in high relative humidity was improved. Recurrence plots and parameter quantification show the variances in the surface corrosion dynamics.

Keywords: cast iron; Paraloid; natural inhibitor; electrochemical noise; recurrence plots

---

1. Introduction

Temporary protection systems based on coatings are usually used in iron objects preserved in museums. The protective layer insulates metal surfaces from moisture, air pollutants, acids, etc. This method provides a passive protection against corrosion [1]. Acrylic systems, such as Paraloid B-72, a copolymer of ethyl methacrylate and methyl methacrylate, have been extensively used for more than twenty years in restoration work as an adhesive and in conservation as a protective film. One of its most notable properties is that, according to Feller [2], Paraloid B-72 is one of the few polymers that has an expected duration of 100 years, and that, under average museum conditions, it can be stored without changes in its transparency. Furthermore, its original solvent is soluble for more than 200 years (according to a projection based on accelerated aging studies). It is a fairly stable resin, nevertheless, its polymer chains deteriorate quickly if exposed to ultraviolet radiation, and its high water permeability decreases its useful life, which is well-known in the industrial field [3].

Another method employed for anticorrosive protection is the use of corrosion inhibitors, which contrast to industrial applications, since, in metal preservation, the addition of the inhibitor to the electrolyte is not mainly used. By definition, inhibitors are used in closed or controlled environments, while objects are exposed to atmospheric conditions that are difficult to control [4]. On the contrary, inhibitors are applied directly on the surface to produce modifications or they are mixed with the varnish. This change in concept and application is important for research on the use of inhibitors in the field of cultural heritage [5–7].

Some inhibitors are being widely used in the preservation and restoration treatments of copper, iron, and silver alloys [8]. For iron alloys, tannic acid and benzotriazole (BTA) are used, tannins being the most used [9]. However, the appearance of a complex of black tannate on the surface is a disadvantage, which is why the use of natural-origin corrosion inhibitors has been proposed [10].

Most of these compounds contain nitrogen, sulfur, and oxygen with a free pair of electrons; additionally, they have aromatic systems [11–13]. These compounds can act on the metal surface through adsorption, by blocking the active sites or by forming a protective layer that can reduce the corrosion rate. N, O, and S are atoms present in the heterocyclic compounds, and are oriented towards the sites where the adsorption probably occurs, because of the availability of a free electron pair [14]. Existing data show that most organic inhibitors act by adsorption at the metal/solution interface, specifically by the displacement of water molecules, which forms a compact film that works as a barrier.

Coating modifications used in the conservation of metallic artifacts are proposed, adding caffeine and nicotine as alternative inhibitors to increase the performance of Paraloid B-72 in high-humidity environments. Improvements are discovered in the mechanism of the metal coating interface through inhibition. Cast iron was selected for evaluation since it shows significant modifications in the corrosion mechanism when atmospheric conditions change. The evaluation of the coatings was carried out through electrochemical noise (EN) with atmospheric corrosion monitors in conditions of 40% and 98% relative humidity (RH), which allows obtaining results in relatively short exposure times due to the high sensitivity of this technique [15–17].

The electrochemical noise technique (the study of the spontaneous fluctuations in voltage and current of the electrode in function of the time) has been employed to determine the anticorrosive resistance in coated substrates [18–22]. This technique allows discerning the corrosion mechanisms that occur through the film, as well as the resistance that it presents. This is why it has proven to be an effective method for comparing coatings performance [23–25]. Time series can be analyzed by statistical and spectral methods, to estimate characteristics that generally describe the behavior of the coated metal [26]. The phenomena that occur on the metal surface in atmospheric conditions are not precisely stationary systems where the dynamics do not change as a function of time. Accordingly, the study of chaotic systems is justified by the existence of many phenomena that have a temporal evolution governed by perfectly deterministic models. An alternative to the statistical analysis of the phenomena that behave in a complex or non-regular way is the recurrence of plots method which allows evaluating the dynamic behavior when identifying the changes in the signals obtained.

The main aim of this research is the study of the current electrochemical noise signal to characterize the dynamics of the corrosion process of the coated iron when relative humidity conditions are modified. Qualitative and quantitative analysis is proposed by recurrence plots (RP) as a tool that promotes the use of the technique to assess coatings, optimizing the information obtained.

2. Materials and Methods

2.1. Materials

Class 25 gray iron was used as working electrode, the chemical composition was set to ensure that variations in composition and microstructure did not alter the electrochemical results. To perform the analysis, the sample was fixed with two parallel surface areas of 5 cm^2. After a suitable preparation, an analysis process was carried out by arc/spark optical emission spectroscopy (OES) analyzer. The equipment used was a SpectroLab Spectrometer brand model LAVWA18B analytical instrument (Mahwah, NJ, USA). The chemical composition obtained is shown in Table 1.

Table 1. Chemical composition determined by an arc/spark OES.

Code	Fe	C	Si	Mn	P	S
Concentration %	92.645	3.65	2.5	0.6	0.9	0.15

The caffeine and nicotine used were Sigma-Aldrich brand (Saint Louis, MO, USA) and were analytical grade reagents with $\geq$99% purity. Paraloid B72 was provided by Carl Roth (Schoemperlenstraße, Karlsruhe, Germany) and the acetone used was provided by J. T. Baker RA (Center Valley, PA, USA). Paraloid B72 was dissolved in acetone 5% (*w/v*) and applied with Kolinsky Brush No. 7, Winsor and Newton (Glasgow, Scotland, UK). To set the concentration on the surface, the Paraloid and the compounds were mixed. For each 100 mL of coating, 5 g of Paraloid B72 and 50 mg of the compound were dissolved in acetone. A total of 0.1 mL of the mixture was applied per square centimeter of area. The inhibitor concentration on the surface was set at 50 $\mu g \cdot cm^{-2}$. Both compounds were added in the preparation to control the amount of inhibitor on the surface, and not on the surface as a pretreatment (as they are most commonly applied). Only one coating layer was applied and placed into a desiccator for 24 h; five linear measurements on the dry film were performed, and the typical standard deviation was calculated. The dry film thickness was set at 30 μm with a deviation of $\pm$5 μm carried out by an Elcometer coating thickness gauge. Measurements were carried out under the ASTM B499-09 Standard [27].

2.2. Sensor Construction (Atmospheric Corrosion Monitor, ACM)

An electrochemical monitor (ACM) was assembled to assess the performance of the coatings in atmospheric conditions. The design consists of an arrangement of three identical electrodes: two electrodes as working electrodes (W_1, W_2) and a central electrode functions as a "pseudo" reference electrode (Ref). The metal plates were cut to 2.5×2.5 cm^2 with a thickness of 0.4 cm. To avoid the short-circuit of the metal electrodes, a 0.1 mm thick Teflon plate was placed between each of them, as shown in Figure 1.

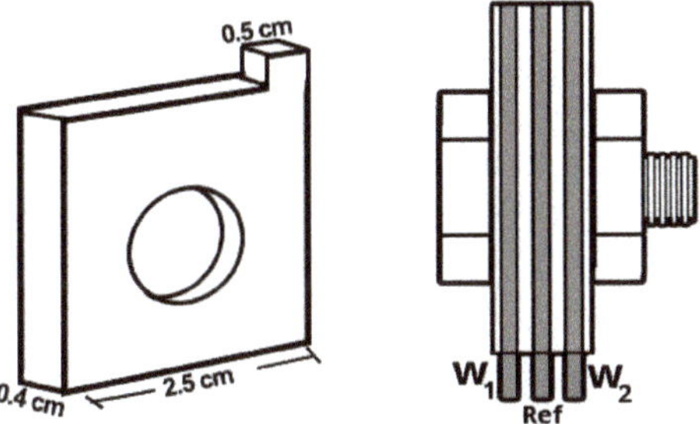

Figure 1. ACM monitors construction.

Each of the plaques was welded to a vulcanized cable, to ensure the continuity of the signal. To maintain the union of the set of plates, Teflon adjusted was used. This arrangement was isolated with epoxy glass resin resistant to chemical attack, as shown in Figure 2.

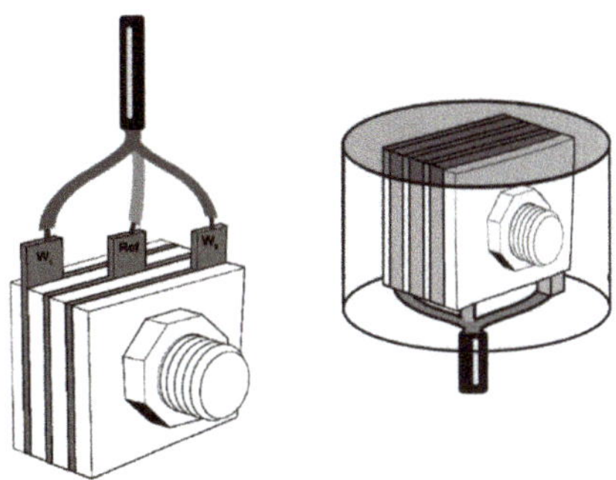

Figure 2. ACM monitor assembled.

Once the monitor is assembled, only a transverse surface was exposed to corrosion. The exposed surface is rough and was polished to a mirror finish. Then it is degreased with acetone and air-dried. This type of corrosion monitor has been applied frequently according to the literature [28,29].

2.3. Relative Humidity

The monitor was placed in a glass desiccator, as shown in Figure 3. The relative humidity was modified with different saline solutions supersaturated under the ASTM E-104 standard [30]. The humidity measures selected during the experimentation phase were 40% RH using potassium carbonate (K_2CO_3) and 98% RH using potassium sulfate (K_2SO_4).

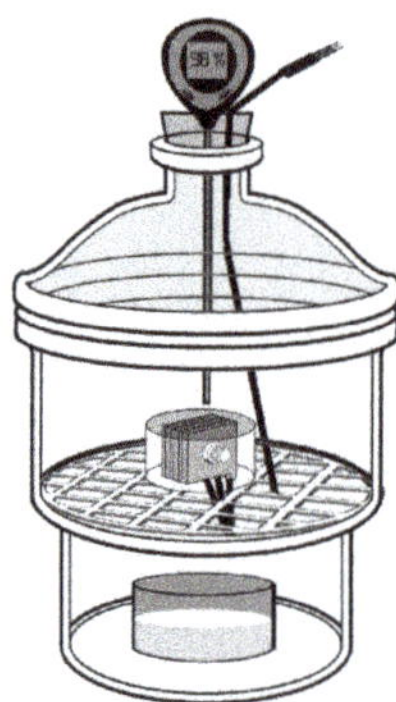

Figure 3. An illustrative diagram of ACM into the desiccator.

The humidity was monitored with a thermo-hygrometer by Instruments, and a thermostat was used to maintain a constant temperature of 25 °C. Two humidity conditions were selected: 40% to evaluate a lower thickness of water on the surface—a percentage of humidity that museums around the world have reported suitable in their exhibitions for the optimal durability of Paraloid and, consequently, the conservation of metal—and 98% relative humidity to have a thickness of water that could be considered as electrochemical corrosion and is the extreme condition.

2.4. Electrochemical Noise Measurements

Voltage and current of electrochemical noise (EN) were measured using a sampling frequency of 0.5 spots per second with 2048 measurements, and a frequency of 1 Hz in order of 0.25 MHz; a measurement was conducted every two hours for 48 h. The measurements were posted out with a zero resistance ammeter (ZRA) by Gill AC 1123 (ACM Instruments, Cumbria, UK); a Faraday cage was used to keep out external interference (static and electromagnetic influences). The resolution was 1 pA for current and 1 µV for voltage measurements. Trend removal was done by subtracting polynomial method of the raw data. The time series were obtained from the response of the system, which allows not only the analysis of the behavior of the inhibition layer but also the determination of its kinetic mechanism. The experiments for each condition were conducted in triplicate to guarantee its reproducibility and reliability.

3. Results

3.1. Electrochemical Noise Analysis

Electrochemical noise signals are registers of current and potential over time. According to this, the random or deterministic behavior of temporal records is conditioned by the possibility of establishing a relationship between the different parameters that govern the corrosion process that is being studied. The comparative analyses of the electrochemical noise signals obtained from a series of potential or current in function of the acquisition time are shown below.

Figure 4 shows the time series of potential and current with the removal of the trend after 24 h of exposure in relative humidity of 40% and 98%. No current density has been shown since the area factor of the electrode is 1 cm^2. To perform a quantitative comparison, the standard deviation of the removed trend values is contained. The blank metal (Figure 4a) to 40% relative humidity presents current and potential lower values (1.8 × 10^{-9} mA, 0.02 mV), compared to the bare metal (Figure 4b) at 98% relative humidity (5.33 × 10^{-7} mA, 0.74 mV). Changes are observed in the fluctuations in both signals obtained as a function of the increase in relative humidity, as a consequence of increasing the thickness of the water film and consequently, the interaction present in the microcells becomes more intense. When the surface is covered with Paraloid (Figure 4c) at 40%, the current and potential changes are not apparently large (1.43 × 10^{-8} mA, 0.23 mV) compared to the bare metal. But for Paraloid (Figure 4d) at 98%, current output decreases (4.43 × 10^{-7} mA) and changes in the amplitude of the potential signal (1.55 mV) are observed, as a consequence of presenting localized active sites where the Paraloid loses protection by absorbing water and the iron corrosion process may occur. When the inhibitors are added to the Paraloid films, modifications are observed as shown in Figure 5.

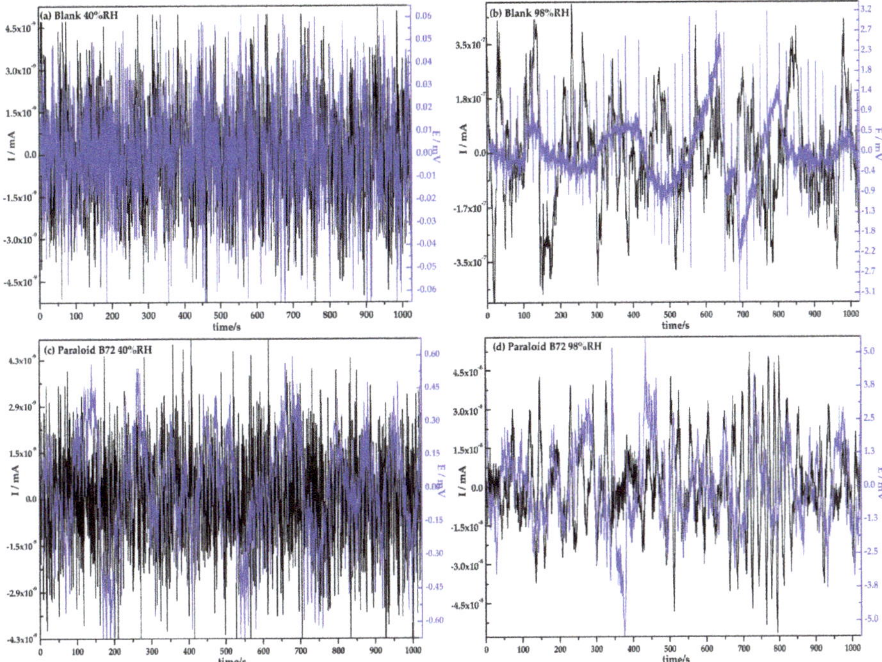

Figure 4. Cast iron's time series in current and potential for (**a**) blank at 40% RH, (**b**) blank at 98% RH, (**c**) Paraloid B72 at 40% RH, and (**d**) Paraloid B72 at 98% RH.

When caffeine is added to Paraloid (Figure 5a) at 40% RH, both current and potential responses decrease (5.06 × 10^{-9} mA, 0.36 mV) and (Figure 5b) for 98% RH (1.75 × 10^{-8} mA, 1.03 mV) almost an order of magnitude, while nicotine films (Figure 5c,d) have smaller values in current and potential at 40% RH (1.69 × 10^{-9} mA, 0.34 mV) as at 98% RH (1.68 × 10^{-9} mA, 0.93 mV) compared to Paraloid without inhibitor. This decrease in the amplitude and response in current and potential reveals the activity of these molecules on the superficial phenomenon of corrosion.

In general, the noise signal apparently has certain changes depending on the conditions, but performing the visual analysis only for a time series is not enough to measure the gain in the impedance of the coating or its performance as a mapping of time.

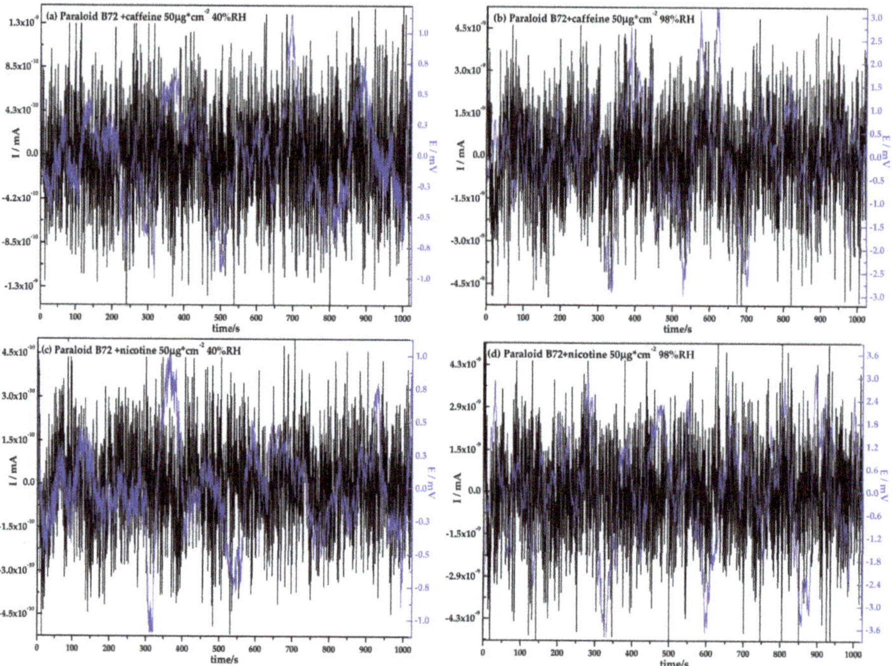

Figure 5. Cast iron's time series in current and potential for (**a**) Paraloid B72 + 50 µg·cm^{-2} with caffeine at 40% RH, (**b**) Paraloid B72 + 50 µg·cm^{-2} with caffeine at 98% RH, (**c**) Paraloid B72 + 50 µg·cm^{-2} with nicotine at 40% of RH, and (**d**) Paraloid B72 + 50 µg·cm^{-2} with nicotine at 98% RH.

This behavior is much more evident when calculating the value of noise resistance (R_n), which is inversely proportional to the intensity of the corrosive attack [31]. Therefore, it is necessary to establish a statistical treatment of the signal to be able to recognize the changes that are submitted in an objective direction and the effect the compounds have along with the Paraloid.

First, the most common statistical parameters, which are the mean and the standard deviation of the signals both in current and in potential, are obtained. Founded on these two statistical parameters, as described in the literature, noise resistance (R_n) is limited. In many investigations, it has been associated with polarization resistance, which is known to be inversely proportional to the corrosion rate when the mechanism is controlled by charge transfer [32]. R_n is defined based on the following Equation (1) [33–35]:

$$R_n = \frac{\sigma_E}{\sigma_I} \tag{1}$$

where σ_I is the standard deviation in current, whereas σ_E is the standard deviation in potential. The second parameter that can be obtained is the so-called localization index (LI), which indicates changes in the corrosion mechanisms, whether homogeneous, mixed or localized, and shows how localized is the attack to the surface of the metal in the medium. This parameter can only take values between 0 and 1, with 0 being a homogeneous attack and 1 being an extremely localized attack [36,37].

The LI is defined based on Equation (2) and only the current noise signal is necessary to determine it. As shown, (X_I) is the value of the mean in current y (σ_I) the standard deviation in current [38]:

$$\text{LI} = \frac{\sigma_I}{\sqrt{\sigma_I^2 + X_I^2}} \tag{2}$$

Figure 6 shows summarizes the statistical analysis of the noise signals of both relative humidity, where both the noise resistance R_n and LI are determined.

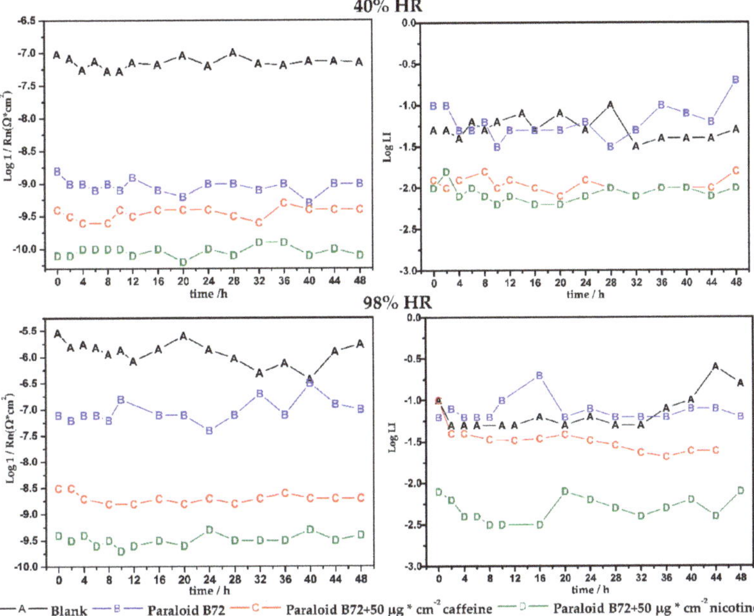

Figure 6. $1/R_n$ (proportional to the corrosion rate) and the location index (LI) according to the times for different coatings in 40% and 98% relative humidity.

Figure 6 shows the statistical analysis of the noise signals in both relative humidities in the function of the exposure time in hours. It is important to note that the logarithm of the reciprocal of R_n is shown to easily appreciate the changes that are generated on the surface with respect to the rate of corrosion. Both R_n and LI are plotted on a logarithmic scale, which works well for comparison purposes, since the differences between the conditions evaluated are very large. In the first place, it is evident that, for both relative humidity, the bare material presents the greatest possible corrosion, as expected. In the relative humidity of 40%, a value of $1/R_n$ is presented in the order of 10^{-7} $(\Omega \cdot cm^2)^{-1}$, while for the relative humidity of 98%, it is 10^{-6} $(\Omega \cdot cm^2)^{-1}$, which is a higher order of magnitude. This result is logical because, in this relative humidity, a continuous water film is formed over the entire surface, resulting in a corrosion mechanism controlled by water and dissolved oxygen, while at 40% there is an ohmic type of control. This clearly shows how humidity is a critical factor in relation to the deterioration of gray iron metal parts. Subsequently, the presence of a physical barrier—the Paraloid B72—clearly decreases the values of $1/R_n$, in the order 10^{-9} $(\Omega \cdot cm^2)^{-1}$ for 40% RH, where it is known that the Paraloid has good performance. However, the Paraloid does not present good protection against high relative humidity, because it tends to allow oxygen and water/moisture, resulting in a value of $1/R_n$ that diminishes two orders of magnitude, up to values of 10^{-7} $(\Omega \cdot cm^2)^{-1}$ for a 98% relative humidity.

This is an important component of this statistical analysis, where it is possible to watch the performance of caffeine and nicotine as corrosion inhibitors. Caffeine has an approximate value of $10^{-9.5}$ $(\Omega \cdot cm^2)^{-1}$ for 40% relative humidity and $10^{-8.5}$ $(\Omega \cdot cm^2)^{-1}$ to 98% relative humidity, while nicotine causes much lower value, of the order of 10^{-10} $(\Omega \cdot cm^2)^{-1}$ to 40% relative humidity and $10^{-9.5}$ $(\Omega \cdot cm^2)^{-1}$ to 98% relative humidity. It is evident how nicotine has a better efficiency as a corrosion inhibitor in both relative humidity.

The location index (LI) also shows notable changes in the different conditions assessed. For a 40% humidity, the average location index of the substrate is $10^{-1.22}$ and for the Paraloid it is $10^{-1.15}$. The LI values in the presence of inhibitor are modified and reach $10^{-1.76}$ with caffeine and $10^{-2.52}$ with nicotine. For a relative humidity of 98%, the trend is modified for bare metal and Paraloid, where the bare substrate has an LI of $10^{-1.03}$ (mixed mechanism), with changes after 32 h at an LI of $10^{-0.76}$, which is identified as a localized mechanism. The change of mechanism is affected by the presence of corrosion products on the surface, which locally changes the reactions. The Paraloid presents an intermediate LI value of $10^{-1.31}$, which represents a mixed mechanism that tends to be localized by the hydration presented by such film. With the addition of caffeine, the LI is approximately $10^{-1.52}$ and with nicotine, it is $10^{-2.04}$.

In other words, when the surface is bare or with Paraloid, the attack mechanism tends to be from mixed to localized as a function of time, while the presence of the inhibitor is manifested by a mixed mechanism for caffeine and a homogeneous one for nicotine. The resistance of Paraloid depends on the relative humidity, which leads to a mixed mechanism for caffeine, presuming that the places where it was not present would be more active due to the water absorption of Paraloid.

It should be noticed that the bare metal has the highest localization index and that the presence of caffeine and nicotine as inhibitors, mainly, reduced the LI value, showing the formation of a homogeneous film that prevents a localized approach. It can also be understood that the proposed used of caffeine and nicotine considerably increases the level of protection and modifies the morphology of the approach.

The purpose of the statistical analysis of the time series is to demonstrate the nature of the corrosion phenomenon. These analysis models are traditionally linear. The ease of the tools is the main line of reasoning in favor of linearity. Yet, theoretically, it is hardly justifiable that this phenomenon of atmospheric corrosion presents a linear behavior. The study of non-linearity is complex, but it is necessary for this type of complex systems. Thus, it is essential to introduce new concepts and instruments referred to non-linearity. Therefore, it is very interesting to be able to define the dynamic execution of this coating, so that in a study of non-linear dynamics the corrosion process is borne out through recursive graphics.

3.2. Recurrence Plots Analysis

Given all the evidence obtained about the chaotic or non-linear behavior, in the series the complete dynamics of the current mechanism cannot be captured through statistical analysis, so the implementation of visual recurrence analysis in complex performances is proposed, so that predictions are allowed for brief periods of time [39]. This method is founded on the immersion theorem of Takens, which establishes that, under certain conditions, it will be possible to receive an estimation that is topologically equivalent, and that it will, therefore, allow extracting all the relevant data about the underlying dynamic system and the unknown that generates the time series. Recurrence plots (RP) are an excellent tool to represent non-linear dynamics and find the repetition of a pattern, although the process is not periodic in the strict sense it is possible to show repetitive or "recurrent" behaviors to differentiate chaotic variability or detect changes in the state of the evolution of a system. Dynamic systems are represented in a phase space that is no more than a vector space, which can have several values. One of the ways to characterize this recurrence is to compare the difference between all the states of the path that describes the evolution of the system that requires obtaining information about the immersion dimension or the dimension of the vector space in which it is possible to represent the dynamics of the system. After choosing a reconstruction dimension and obtaining the vectors, the RP represents a lot of stages in a square of dimension $M \times M$, where M, the axes or the sides of the second power, represent the chronological succession of the vectors in the remodeled space. This analysis was based on the methodology proposed by Eckmann et al. [40]. It consists of specifying when the levels in the reconstructed phase space are infinitesimally close because they have applied a very small dimension. The progressive increase of the dimension until the false infinitesimals disappear provides a criterion of the necessary dimension for the reconstruction.

Subsequent developments have allowed quantifying the amount of recurrence present in the graph. Zbilut and Webber [41] propose the RQA (recurrence quantification analysis) as defined by the following characteristic indices:

The percentage of recurrence (%Rec) consists of the percentage of points that are in the value threshold, or the lowest value in the total points in the recursive graph. This parameter is associated with the periodicity of the signal. Described in Equation (3) as:

$$\%\text{Rec} = \frac{\text{NRECURS}/M(M-1)}{2} \cdot 100 \tag{3}$$

NRECURS is the entire number of recurring periods in the superior triangle of the graph without counting the peaks of the bisector. The denominator of Equation (3) is the number of points in the upper triangle of the graph eliminated those of the bisector.

The percentage of determinism (%Det) is the percentage of stops that constitute a line parallel to the main diagonal. This parameter is a measurement of how much the events in the past affect those in the future. Described in Equation (4) as:

$$\%\text{Det} = \frac{\text{DRECURS}}{\text{NRECURS}} \cdot 100 \tag{4}$$

DRECURS is the number of points that are part of line segments parallel to the bisector of the square. A line segment is specified as two or more adjacent points.

The maximum line (LM) is the duration of the most recurrent segment and is reciprocally proportional to the maximum Lyapunov exponent, which, definitively, tells us about the sensibility of our system to the initial conditions. Finally, the entropy of information or Shannon Entropy corresponds to the probabilistic distribution of the diagonal lines of determinism and is related to the complexity of the determinant structure in the system [42]. Described in Equation (5) as:

$$\text{Entropy} = -\sum_{i=0}^{n} p_i \, \log p_i \tag{5}$$

where p_i is the relative frequency of the length of the recurring segments. If the series is random, this measure is 0.

The analysis mutual information (AMI) function was used to establish the delay time, and the false nearest neighbors (FNN) method was used for the immersion dimension to obtain the reconstruction parameters through the method proposed by Dämming and Mitschke [43]. The analysis of recurrence values was performed by the method proposed by Garcia-Ochoa et al. [44].

The RP obtained for the 24-h exposure time series are represented below in all the conditions assessed for the current time series, since they are directly linked to the localization index of the statistical analysis. As mentioned, the most important changes in signal amplitude are shown.

Figure 7 shows the graphs of recurrence for all the conditions assessed. Recurrence plots represent the dynamics in the time series in a two-dimensional space whose axes represent the number of data in the series (2048 data). The presence of regularities obtained in the dynamics of the time series appears in the graph. Time-series are sets of swings that have some periodicity and therefore, RP presented lines and correlation structures, which is suggestive of a stochastic behavior [45,46].

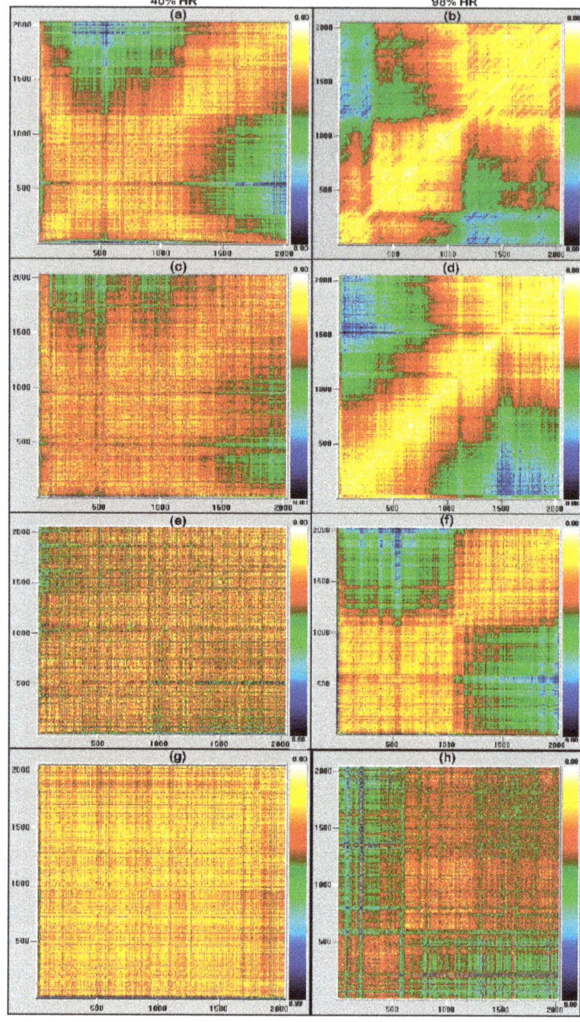

Figure 7. Recurrence plots of (**a**) Blank at 40% RH, (**b**) blank at 98% RH, (**c**) Paraloid B72 at 40% RH, (**d**) Paraloid B72 at 98% RH, (**e**) Paraloid B72 + μg·cm^{-2} of caffeine at 40% RH, (**f**) Paraloid B72 + 50 μg·cm^{-2} of caffeine 98% at RH, (**g**) Paraloid B72 + 50 μg·cm^{-2} of nicotine at 40% RH, and (**h**) Paraloid B72 + 50 μg·cm^{-2} of nicotine at 98% RH.

In the case of chaotic dynamics, short lines parallel to the main diagonal appear, while in the case of a random dynamics, the plot shows a uniform representation of points indicating that there is some structure in the data. In 40% relative humidity, the signals observed present more organized graphs, which shows a weakly consolidated portion of parallel structures, indicating periodicity in comparison to signals in 98% relative humidity. Furthermore, it was noted how the dynamics of the system contracts in some spots and then expands: this is a feature of systems with chaotic dynamics.

When the metal is coated with Paraloid, the formation of equidistant geometric structures was observed, which is presumed to be a more deterministic behavior. However, in this immersion condition, the recurrence plot shows a portion of reinforced parallel structures. The presence of caffeine and nicotine in two humidity conditions assessed features more structured graphics in comparison to the Paraloid, which will be reflected on signals with a higher degree of recurrence and will manifest themselves in a greater number of yellow and reddish spots. To this point, the analysis of graphs has been qualitative, and presumptive behavior may be viewed. To perform a more detailed analysis, we calculated the percentage of recurrence, the percentage of determinism, entropy and maximum line depending on weather for all series in the conditions evaluated at 40% and 98% relative humidity, set by the Euclidean method. The results are shown below in Figure 8.

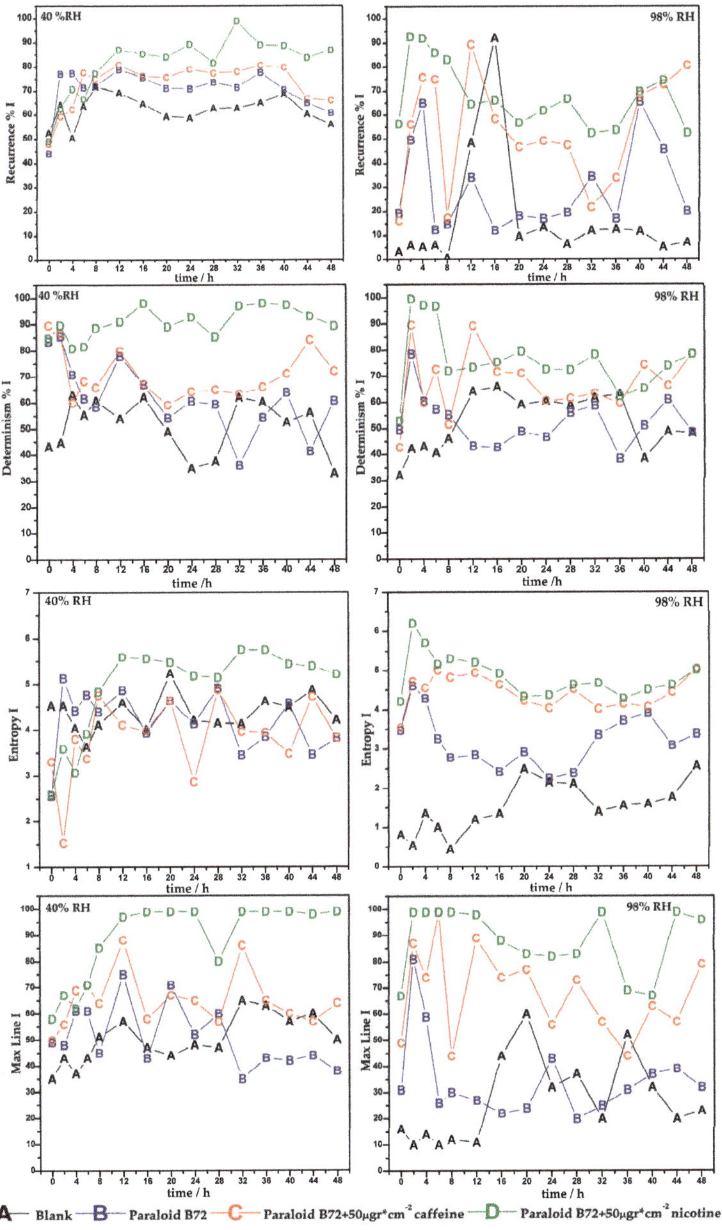

Figure 8. Recurrence time as a function of time in 40% and 98% relative humidity.

The percentage of recurrence (%Rec) presents greater sensitivity, showing drastic changes as a mapping of relative humidity for uncoated iron, showing a lower value for bare metal at 98% relative humidity. This confirmation that iron modifies its periodic behavior when in contact with water, so the mechanisms that have been identified in the literature are corroborated through the recurrence of the signal. The Paraloid also presents modifications with respect to humidity, since in a relative humidity of 98% the system begins to be more random: this means that the system variables change in function of time, and then that the Paraloid, when in the presence of water, loses its protective capacities and limits its use because it will fail randomly.

The presence of nicotine in both humidity conditions reaches the highest recurrence values revealing a more periodic signal, this being a first evidence of the effect on system dynamics when adding the compound. The quantity of interaction between the different sites where the corrosion takes place, which corresponds to the percentage of determinism (%Det), is shown below. This parameter is remarkably sensitive to the surface conditions and the interaction of the microcells. First, %Det is minor for bare metal. This implies great randomness in the interactions of the electrochemical cells, since, for the case of a 40% relative humidity, an average value of 40% Det is initiated, while at a humidity of 98%, it is 40% Det during the first 10 h to values of 60% Det after 12 h, which describes how the interactions are modified when the surface presents corrosion products. This converts the response into a more synchronized system.

When the material is covered by the Paraloid, %Det is increased substantially to values of 80% at 40% relative humidity, and 50% at 98% relative humidity, but it changes with the same tendency as the percentage of recurrence, which decreases in function of time, showing again how the Paraloid loses protection in function of time, which decreases the synchronization and increases the interaction areas, allowing the surface to present arbitrary responses. When adding caffeine or nicotine, the remarkable growth of the %Det degree is quite noticeable in both humidity, this being more remarkable for the relative humidity of 98%, which, as cited above, implies the shaping of a water film on the airfoil. It can be seen that the chemical construction of both substances intervenes remarkably in the kinetics of the electrochemical microcells, being nicotine both the one that reported a greater degree of determinism and the one that reported a lower level of corrosion and localization.

It can be noted that the value of the entropy of information increases remarkably with the bearing of both caffeine and nicotine, the latter giving the highest value reported for both relative humidity. This is indicative of the increment in the complexity of the corrosion process dynamics that takes place, showing that by increasing the level of determinism, the level of complexity increases too, hence yielding a higher degree of security for both uniform and localized corrosion. Then it could be stated that there is a procedure of self-formation that results in greater protection.

Finally, the maximum line, which is an index of the level of sensibility to the initial conditions, once again demonstrates that nicotine and caffeine cause the participating sites to interact so that the system shows greater synchronization and, as a consequence, the most positive coefficient of Lyapunov is smaller.

4. Discussion

Significant changes are noted in the percentage of determinism since this is modified regarding the protection applied and the percentage of humidity assessed. The metal in different humidities has a less synchronized behavior. The best performance in anticorrosive protection is in Paraloid with nicotine, followed by Paraloid with caffeine. A relation based on the signals obtained and, on the parameters, analyzed shows that %Det indicates protection and continuity of the inhibition of the coatings over time. The dynamics of the systems have been confirmed as chaotic for both the uncoated metal and Paraloid B72 in both relative humidities since it was found that the entropy of both systems decreases, which shows that the systems have a sensitive dependence to the initial conditions.

Overall, it could be said that, based on the analysis of recursive graphs, the presence of caffeine and nicotine decreases the possibility of Paraloid to capture water and, thus, the metallic material will have fewer interactions. The inhibitors are taken up locally at these sites, which is in complete accord with the statistical analysis of the electrochemical noise signal. In this mode, the behavior of the coating is modified by adding both caffeine and nicotine, modifying and increasing the resistance when conferring protection, because when these compounds are added, the kinetics of the corrosion process are modified. This allows for the introduction of a system that is self-organized as a strategy to use the Paraloid, as the temporary security for its useful life is guaranteed in conditions where the relative humidity cannot be manipulated.

5. Conclusions

Electrochemical noise is capable to assess with great sensitivity the increase in corrosion resistance performance conferred by the inhibitors added to Paraloid; even the electrolytic resistance changes corrosion mechanism.

The addition of nicotine and caffeine to temporary protection is contemplated in a higher protection efficacy regardless of the relative humidity assessed. Nicotine presented a high degree of protection, superior to the caffeine protection added to Paraloid coatings.

The nonlinear analysis of the electrochemical noise signal by recurrence plots shows the variances in the surface corrosion dynamics, where a more synchronized process resulting in greater protection.

Author Contributions: This paper was written by all authors. P.R.B. and F.-J.R.G. conceived and designed the experiments; P.R.B. conducted a research and investigation process, specifically performed the experiments; P.R.B., F.-J.R.G. and E.G.O. discussed the results and revised the paper; P.R.B. wrote the paper.

Funding: Thanks to CONACYT for the financial support for the development of this research through the basic science project 239938.

Acknowledgments: Thanks to CONACYT for the scholarship granted to Paola Roncagliolo Barrera with (Scholarship CVU number: 332740) to develop her PhD research. The authors wish to thank to Carlos Rodríguez Rivera for the experimental assistance provided as technical supervisor throughout this study.

Conflicts of Interest: The authors declare no conflict of interest.

References

1. Caple, C. *Conservation Skills: Judgement, Method and Decision Making*; Routledge: Abingdon, UK, 2012.
2. Feller, R.L. Aspects of chemical research in conservation: The deterioration process. *J. Am. Inst. Conserv.* **1994**, *33*, 91–99. [CrossRef]
3. González, M.L.G.; Gomez, M.L. *La Restauración: Examen Científico Aplicado a la Conservación de Obras de arte*; Cátedra: Madrid, Spain, 1998. (In Spanish)
4. Garcia, E.; Díaz, S. *Técnicas Metodológicas Aplicadas a la Conservación-Restauración del Patrimonio Metálico*; Ministerio de Cultura: Madrid, Spain, 2011. (In Spanish)
5. Mirambet, F.; Reguer, S.; Rocca, E.; Hollner, S.; Testemale, D. A complementary set of electrochemical and X-ray synchrotron techniques to determine the passivation mechanism of iron treated in a new corrosion inhibitor solution specifically developed for the preservation of metallic artefacts. *Appl. Phys. A* **2010**, *99*, 341–349. [CrossRef]
6. Dillmann, P.; Beranger, G.; Piccardo, P.; Matthiessen, H. *Corrosion of Metallic Heritage Artefacts: Investigation, Conservation and Prediction of Long Term Behaviour*; Elsevier: New York, NY, USA, 2014.
7. Cano, E.; Lafuente, D. Corrosion inhibitors for the preservation of metallic heritage artefacts. In *Corrosion and Conservation of Cultural Heritage Metallic Artefacts*; Elsevier: New York, NY, USA, 2013; pp. 570–594.
8. Liu, A.M.; Ren, X.F.; Wang, B.; Zhang, J.; Yang, P.X.; Zhang, J.Q.; An, M.Z. Complexing agent study via computational chemistry for environmentally friendly silver electrodeposition and the application of a silver deposit. *RSC Adv.* **2014**, *4*, 40930–40940. [CrossRef]
9. Hollner, S.; Mirambet, F.; Rocca, E.; Reguer, S. Evaluation of new non-toxic corrosion inhibitors for conservation of iron artefacts. *Corros. Eng. Sci. Technol.* **2010**, *45*, 362–366. [CrossRef]

10. Cano, E.; Bastidas, D.M.; Argyropoulos, V.; Fajardo, S.; Siatou, A.; Bastidas, J.; Degrigny, C. Electrochemical characterization of organic coatings for protection of historic steel artefacts. *J. Solid State Electr.* **2010**, *14*, 453. [CrossRef]

11. Satri, V. *Green Corrosion Inhibitors: Theory and Practice*; John Wiley & Sons: Hoboken, NJ, USA, 2011.

12. Rahmouni, K.; Takenouti, H. Struggle against corrosion: Protection by triazoles compounds of ancient and modern bronzes covered with patina. *Actual. Chim.* **2009**, *327–328*, 38–44.

13. Al-Otaibi, M.S.; Al-Mayouf, A.M.; Khan, M.; Mousa, A.A.; Al-Mazroa, S.A.; Alkhathlan, H.Z. Corrosion inhibitory action of some plant extracts on the corrosion of mild steel in acidic media. *Arab. J. Chem.* **2014**, *7*, 340–346. [CrossRef]

14. Okafor, P.; Ikpi, M.; Uwaha, I.; Ebenso, E.; Ekpe, E.; Umoren, S. Inhibitory action of Phyllanthus amarus extracts on the corrosion of mild steel in acidic media. *Corros. Sci.* **2008**, *50*, 2310–2317. [CrossRef]

15. Zajec, B.; Leban, M.B.; Lenart, S.; Gavin, K.; Legat, A. Electrochemical impedance and electrical resistance sensors for the evaluation of anticorrosive coating degradation. *Corros. Rev.* **2017**, *35*, 65–74. [CrossRef]

16. Xia, D.H.; Ma, C.; Song, S.Z.; Ma, L.L.; Wang, J.H.; Gao, Z.M.; Zhong, C.; Hu, W.B. Assessing atmospheric corrosion of metals by a novel electrochemical sensor combining with a thin insulating net using electrochemical noise technique. *Sens. Actuators B Chem.* **2017**, *252*, 353–358. [CrossRef]

17. Ma, C.; Song, S.Z.; Gao, Z.M.; Wang, J.H.; Hu, W.B.; Behnamian, Y.; Xia, D.H. Electrochemical noise monitoring of the atmospheric corrosion of steels: Identifying corrosion form using wavelet analysis. *Corros. Eng. Sci. Technol.* **2017**, *52*, 432–440. [CrossRef]

18. Schaefer, K.; Mills, D.J. The application of organic coatings in conservation of archaeological objects excavated from the sea. *Prog. Org. Coat.* **2017**, *102*, 99–106. [CrossRef]

19. Bierwagen, G.P.; Wang, X.; Tallman, D.E. In situ study of coatings using embedded electrodes for ENM measurements. *Prog. Org. Coat.* **2003**, *46*, 163–175. [CrossRef]

20. Puget, Y.; Trethewey, K.; Wood, R.J.K. Electrochemical noise analysis of polyurethane-coated steel subjected to erosion-corrosion. *Wear* **1999**, *233*, 552–567. [CrossRef]

21. Mansfeld, F.; Han, L.T.; Lee, C.C.; Chen, C.; Zhang, G.; Xiao, H. Analysis of electrochemical impedance and noise data for polymer coated metals. *Corros. Sci.* **1997**, *39*, 255–279. [CrossRef]

22. Skerry, B.S.; Eden, D.A. Characterization of coatings performance using electrochemical noise-analysis. *Prog. Org. Coat.* **1991**, *19*, 379–396. [CrossRef]

23. Greisiger, H.; Schauer, T. On the interpretation of the electrochemical noise data for coatings. *Prog. Org. Coat.* **2000**, *39*, 31–36. [CrossRef]

24. Kiele, E.; Lukseniene, J.; Griguceviciene, A.; Selskis, A.; Senvaitiene, J.; Ramanauskas, R.; Raudonis, R.; Kareiva, A. Methyl-modified hybrid organic-inorganic coatings for the conservation of copper. *J. Cult. Herit.* **2014**, *15*, 242–249. [CrossRef]

25. Mohamed, W.A.; Mohamed, N.M. Testing coatings for enameled metal artifacts. *Int. J. Conserv. Sci.* **2017**, *8*, 15–24.

26. Mills, D.; Picton, P.; Mularczyk, L. Developments in the electrochemical noise method (ENM) to make it more practical for assessment of anti-corrosive coatings. *Electrochim. Acta* **2014**, *124*, 199–205. [CrossRef]

27. *ASTM b499–09 Standard Test Method for Measurement of Coating Thicknesses by the Magnetic Method: Nonmagnetic Coatings on Magnetic Basis Metals*; ASTM International: West Conshohocken, PA, USA, 2014.

28. Gonzalez, J.A.; Otero, E.; Cabanas, C.; Bastidas, J.M. Electrochemical sensors for atmospheric corrosion rates—A new design. *Br. Corros. J.* **1984**, *19*, 89–94. [CrossRef]

29. Lecuyer, C.; Barreau, C.; Thierry, D. Electrochemical sensor for in-situ monitoring of coated metals degradation under atmospheric conditions. *Mem. Etud. Sci. Rev. Met.* **1991**, *88*, 691–701.

30. *ASTM e104–02 Standard Practice for Maintaining Constant Relative Humidity by Means of Aqueous Solutions*; ASTM International: West Conshohocken, PA, USA, 2012.

31. Sanchez-Amaya, J.M.; Cottis, R.A.; Botana, F.J. Shot noise and statistical parameters for the estimation of corrosion mechanisms. *Corros. Sci.* **2005**, *47*, 3280–3299. [CrossRef]

32. Bertocci, U.; Gabrielli, C.; Huet, F.; Keddam, M. Noise resistance applied to corrosion measurements I. Theoretical analysis. *J. Electrochem. Soc.* **1997**, *144*, 31–37. [CrossRef]

33. Kearns, J.R.; Metals, A.C.G. *Electrochemical Noise Measurement for Corrosion Applications*; ASTM: West Conshohocken, PA, USA, 1996.

34. Cottis, R.; Turgoose, S. *Electrochemical Impedance and Noise*; NACE International: Houston, TX, USA, 1999.

35. Mabbutt, S.; Mills, D.J.; Woodcock, C.P. Developments of the electrochemical noise method (ENM) for more practical assessment of anti-corrosion coatings. *Prog. Org. Coat.* **2007**, *59*, 192–196. [CrossRef]

36. Mansfeld, F.; Sun, Z. Localization index obtained from electrochemical noise analysis. *Corrosion* **1999**, *55*, 915–918. [CrossRef]

37. Cottis, R.A.; Al-Awadhi, M.A.A.; Al-Mazeedi, H.; Turgoose, S. Measures for the detection of localized corrosion with electrochemical noise. *Electrochim. Acta* **2001**, *46*, 3665–3674. [CrossRef]

38. Mansfeld, F. The electrochemical noise technique—Applications in corrosion research. *AIP Conf. Proc.* **2005**, *780*, 625–630.

39. Pham, T.D. From fuzzy recurrence plots to scalable recurrence networks of time series. *EPL-Europhys. Lett.* **2017**, *118*, 20003. [CrossRef]

40. Eckmann, J.-P.; Kamphorst, S.O.; Ruelle, D. Recurrence plots of dynamical systems. *EPL-Europhys. Lett.* **1987**, *4*, 973. [CrossRef]

41. Zbilut, J.P.; Webber, C.L. Embeddings and delays as derived from quantification of recurrence plots. *Phys. Lett. A* **1992**, *171*, 199–203. [CrossRef]

42. Trulla, L.L.; Giuliani, A.; Zbilut, J.P.; Webber, C.L. Recurrence quantification analysis of the logistic equation with transients. *Phys. Lett. A* **1996**, *223*, 255–260. [CrossRef]

43. Mitschke, F.; Dämmig, M. Chaos versus noise in experimental data. *Int. J. Bifurcat. Chaos* **1993**, *3*, 693–702. [CrossRef]

44. Cazares-Ibanez, E.; Vazquez-Coutino, G.A.; Garcia-Ochoa, E. Application of recurrence plots as a new tool in the analysis of electrochemical oscillations of copper. *J. Electroanal. Chem.* **2005**, *583*, 17–33. [CrossRef]

45. Webber, C.L., Jr.; Ioana, C.; Marwan, N. *Recurrence Plots and Their Quantifications: Expanding Horizons*; Springer International Publishing: Cham, Switzerland, 2016.

46. Webber, C.L., Jr.; Marwan, N. *Recurrence Quantification Analysis: Theory and Best Practices*; Springer: Berlin, Germany, 2014.

Article

Determination of Optimum Concentration of Benzimidazole Improving the Cathodic Disbonding Resistance of Epoxy Coating

Saghar Nabavian [1], Reza Naderi [2,*] and Najmeh Asadi [2]

[1] Department of Marine Science and Technology, College of Marine Chemistry, Islamic Azad University, P.O. Box 14515-755 Tehran, Iran; sagharnabavian@gmail.com

[2] School of Metallurgy and Materials Engineering, College of Engineering, University of Tehran, P.O. Box 11155-4563 Tehran, Iran; najmehasadi@ut.ac.ir

* Correspondence: rezanaderi@ut.ac.ir; Tel.: +98-21-82084075; Fax: +98-21-88006076

Received: 7 November 2018; Accepted: 13 December 2018; Published: 19 December 2018

Abstract: This study is aimed to evaluate the effect of concentrated benzimidazole (BIM) on the cathodic disbonding (CP) of an epoxy coating applied on steel substrate. For this purpose, the polymeric coatings, formulated with different concentrations of BIM (0 wt.%, 0.5 wt.%, 0.75 wt.%, and 1 wt.%, were subjected to the CP test at the potential of -1.2 V vs. Ag/AgCl during 24 h immersion in 3.5 wt.% NaCl solutions. The optimum formulation was found through taking advantage of the CP test results, FESEM/EDX, and EIS data. Moreover, a pull-off test was used to measure the wet adhesion strength. For insight into the inhibition function of the organic inhibitor, the behavior of steel in the sodium chloride solutions, with and without BIM, was compared using EIS and surface analysis.

Keywords: epoxy coating; EIS; organic inhibitor; benzimidazole; cathodic disbanding

1. Introduction

For protection of metals exposed to a corrosive environment, application of organic coatings is a reliable way. In the case of coated metals, bare metal is exposed to the aggressive environment in the damaged or defect areas of coating, leading to the occurrence and progress of underfilm corrosion. Therefore, cathodic protection is proposed to resolve the problem. In fact, organic coatings applied on the metallic substrates reduce the current required for cathodic protection. Despite the many advantages of cathodic protection, the technique may increase the possibility of coating delamination. Some parameters such as coating thickness, coating composition, surface treatment, dissolved oxygen, NaCl concentration, and cathodic potential can affect the delaminated area [1–6]. The voltage applied for cathodic protection promotes the reduction of oxygen penetrating through the coating matrix. Therefore, OH^- ions accumulate in the metal/coating interface. As sodium cations access the disbonding front, the alkalinity can destroy the bond between the coating and metal surface, particularly on the edge of defects [7–14].

One of the effective strategies for controlling the film delamination is to modify the coating formulation. For instance, addition of anticorrosion pigments [15–17], organic inhibitors, e.g., azole derivatives [18–20], thioglycolate esters, mercaptocarboxylic acids [21], organic sulfides, organic amines, organic phosphates, phenols [22] disodium oleamidesulfosucinate, lignosulfonic acid-doped polyanilin [23], and also inhibitor loaded nanocontainers [24] have been proposed to increase the coating resistance against the destroying phenomena. The electrolyte penetrating into the coating may transfer organic inhibitor molecules and inhibiting species liberated from pigment particles to the interface, where the materials make a complex passive layer on the metal surface, blocking the active regions and reducing the rate of electrochemical reactions [25–29]. In the case of inhibitor loaded nanocontainers, Izadi et al. [24], for example, showed that release of Nettle molecules as green corrosion inhibitor and zinc cations from a nanocontainer, synthesized by l-b-l process with the core of Fe_3O_4 nanoparticles, provide epoxy coating with a significant cathodic disbonding protection. Moreover, the positive effect of zinc aluminum hydrotalcite intercalated with benzothiazolylthio-succinic acid on the cathodic delamination resistance of epoxy coating was reported by Hang et al. [30]. Interestingly, the influence of either direct or indirect addition of corrosion inhibitors to the conversion coatings or any pretreatments on the cathodic delamination of overlaying polymeric films has been studied in some literature [24,31,32]. Ramezanzadeh et al. [31] achieved an improved cathodic disbonding and adhesion properties through application of epoxy coating on the steel substrate covered by zinc phosphate conversion coating containing poly(vinyl) alcohol. Pretreatment of steel with silane layer incorporating amino and isocyanate silane functionalized graphene oxide nanosheets was shown to decrease the cathodic disbondment of epoxy top coating [32].

Previously, we reported the positive role of second and third generations of phosphate based anticorrosion pigments on the function of epoxy coating on the steel substrate in the condition of its application of cathodic protection [15–17]. In other words, zinc aluminum phosphate, zinc aluminum polyphosphate, and strontium aluminum polyphosphate (zinc-free pigment) were shown to play an important role in the disbonding front through decreasing pH and the deposition of an insoluble layer on the surface, which disrupts electrochemical reactions and increases the coating-metal bonding strength. In this work, we include benzimidazole as an organic corrosion inhibitor to an epoxy coating formulation to reduce the cathodic disbandment rate. In order to find the most effective inhibitor concentration, electrochemical impedance spectroscopy (EIS), pull-off test, and surface analysis methods were employed.

2. Experimental

2.1. Materials and Sample Preparation

ST37 steel panels with the composition mentioned in the previous work [15] and with a 1 mm thickness were polished using magnetic polisher to reach a desirable surface, then degreasing with acetone. In the solution phase study, the panels were immersed in 3.5 wt.% NaCl aqueous solutions containing 1 mM of Benzimidazole (BIM), purchased from Merck and used with no further purification. To prepare polymeric coatings, 0 wt.%, 0.5 wt.%, 0.75 wt.%, and 1 wt.% of the organic inhibitor were mixed in polyamide (Crayamid 115). Then, the prepared polyamide was mixed in epoxy resin Epiran-01X75 (Khouzestan petrochemical Co., Bandar Jomeiny, Iran) at a stoichiometric ratio of 60:100. BYK-306 (BYK-Chemie GmbH, Wesel, Germany) was used as a leveling agent in the coating formulation. After curing at a temperature of 80° C for 40 min, the thickness of dry film measured using Elcometere-445 (Elcometer Inc., Manchester, UK) was approximately 20 ± 3 μm.

2.2. Methods

The electrochemical impedance spectroscopy was carried out using Autolab PGSTAT12 (Metrohm AG, Herisau, Switzerland) an open circuit potential (OCP) in the frequency domain of 10 kHz to 10 mHz using a 10 mV amplitude perturbation. A setup containing the sample (bare metal or coated substrate) as working electrode, platinum counter electrode, and Ag/AgCl reference electrode was used for EIS test. The electrical connection was provided by a copper wire attached to one surface of the working electrode. For the bare plates with 1 cm^2 surface area exposed to 3.5 wt.% NaCl solutions with and without 1 mM benzimidazole, EIS tests were performed at room temperature after 5 and 24 h of immersion. The electrochemical tests for the coated samples after subjection to a cathodic disbonding test were conducted after 12 and 24 h immersion in 3.5 wt.% NaCl solution. The data of EIS performed on three replicates was analyzed with the use of ZsimpWin software (v3.22).

The steel samples, which were covered by the epoxy coatings containing different amounts of BIM with an artificial circular hole (1 mm in diameter), were subjected to cathodic disbonding tests. In the test, the samples immersed in 3.5 wt.% NaCl solution were polarized at −1.2 V vs. Ag/AgCl. Some radial cuts intersecting at the hole were provided on the films at the end of cathodic disbonding test and the detached coatings were removed with a sharp knife to determine the average surface area of disbondment. In other words, the average distance between the hole and intact parts of polymeric film was considered as the radius of disbondment.

After attachment of dollies on the epoxy coating using a 2-part Araldit epoxy, a pull-off test was performed with a PosiTest digital adhesion tester (DeFelsko Corp., Ogdensburg, NY, USA). In the wet mode, the dollies were detached from the sample surface after 24 h immersion in 3.5 wt.% NaCl solution to obtain bonding strength of the coatings with and without the organic inhibitor.

A field emission electron microscopy (FE-SEM/EDS, TESCAN MIRA 3 Lumbers, Brno, Czech Republic) was employed to characterize the surface of (1) bare steel after 24 h dipping in 3.5 wt.% NaCl solution with 1 mM BIM and (2) areas beneath the delaminated epoxy coating with BIM after a cathodic disbonding test.

3. Results and Discussion

Prior to introducing of the inhibitor into the epoxy coating, the effect of BIM on the corrosion of uncoated ST37 panels in the sodium chloride electrolyte was evaluated by the electrochemical impedance spectroscopy technique. Figure 1 compares the Nyquist plots of bare specimens after five and 24 h immersion in 3.5 wt.% NaCl electrolyte with and without 1 mM BIM. In the presence of the organic molecule in the two immersion periods, semicircles with larger diameter were visible, indicating a kind of corrosion inhibition. Moreover, increasing the time of exposure to the blank electrolyte resulted in smaller spectra, while the exposure period had no noticeable effect on the spectra in the presence of BIM. Since only one relaxation time was detected in the AC impedance spectra of both cases after five and 24 h exposure, the simple equivalent circuit illustrated in Figure 2 was selected to model the data. As shown in Figure 2, R_s represents the solution resistance, R_{ct} the charge transfer resistance, and CPE_{dl} the constant phase element of double layer. The data resulting from the modeling is presented in Table 1.

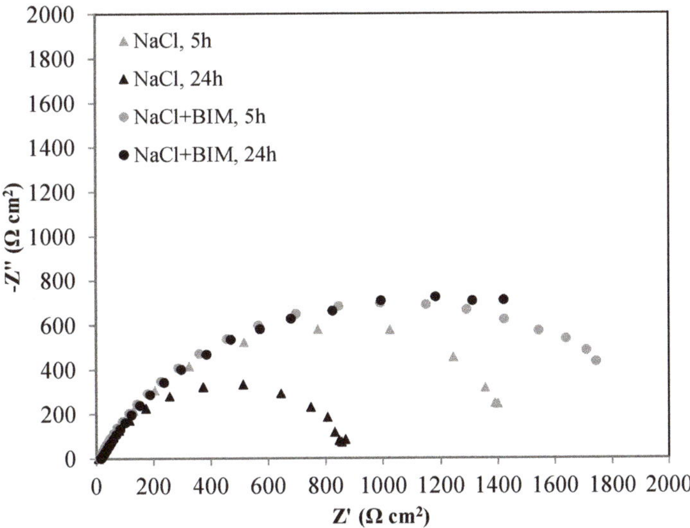

Figure 1. Nyquist diagrams of the bare specimens dipped in 3.5 wt.% NaCl electrolyte with and without 1 mM BIM after 5 and 24 h.

In order to determine the double layer capacitance values in Table 1, Equation (1) was used [33]:

$$C_{dl} = (Y_0 R_{ct}^{-n})^{1/n} \tag{1}$$

where Y_0 and n are, respectively, the admittance and exponent of CPE.

Some important features can be derived from the table. The presence of BIM in the aggressive electrolyte resulted in a decrease in double layer capacitance and increase in the charge transfer resistance. For instance, the R_{ct} value in the presence of BIM was approximately 2.5 times higher than that in the absence of BIM at the end of the 24 h dipping period. Besides the magnitude, increasing the time of exposure to the blank sodium chloride solution had a decreasing effect on the charge transfer resistance, while an increasing trend was observed for the metallic specimen dipped in the electrolyte with BIM. Considering the inverse proportion of the charge transfer resistance to corrosion current density, the trend and magnitude of the resistance parameter in the presence of the organic molecule indicated an inhibition function [34]. The replacement of a water molecule with the organic molecules (BIM) might be responsible for the drop in the double layer capacitance values after both five and 24 dipping durations. These results were in agreement with the data we acquired for some organic inhibitors in the chloride solution [35,36]. The organic molecules existing in the extract of Mentha longifolia in sodium chloride solution was reported to provide a maximum inhibition efficiency of 69.5% after 24 h [35]. We also found, in another research, that the presence of 4,5-imidazoledicarboxylic in NaCl solution led to a kind of corrosion inhibition for steel plates and the obtained impedance spectra during 24 h immersion had only a one time constant [36].

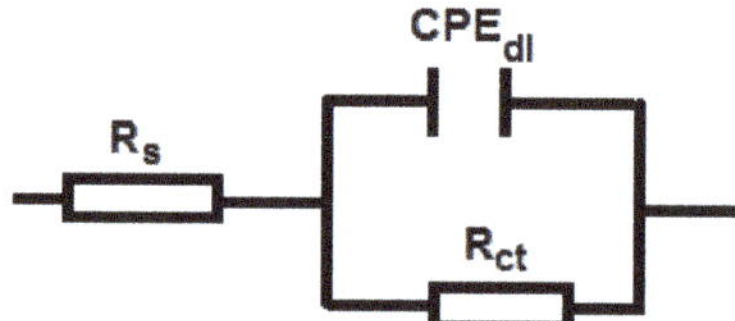

Figure 2. The equivalent circuit for modeling of the results of EIS test on the bare ST37 steels after 5 and 24 h dipping in 3.5 wt.% NaCl solution in the absence and presence of 1 mM BIM.

Table 1. The data resulting from modeling of the AC impedance spectra of bare ST37 steels after 5 and 24 h dipping in 3.5 wt.% NaCl solution in the absence and presence of 1 mM BIM.

BIM Concentration (mM)	Immersion Time (h)	R_{ct} ($\Omega\cdot$cm^2)	C_{dl} (μF$\cdot$cm^{-2})
0	5	1292	752
	24	912	529
1	5	2001	469
	24	2256	404
Standard deviation	R_c		1.7%–9.3%
	C_{dl}		3.8%–13.8%

The appearance of bare ST37 specimens after 24 h immersion in the NaCl solution with and without the organic moiety is shown in Figure 3. A decrease of corrosion products on the specimen surface immersed in the electrolyte containing benzimidazole was clearly visible, confirming the electrochemical data.

In order to further investigate the function of BIM, SEM-EDS tests were performed on the surface of bare ST37 specimens after 24 h immersion in the electrolyte (Figure 4). As seen from Figure 4, a layer was observed on the surface exposed to the NaCl solution containing BIM. The elemental composition of the detected layer, which was obtained by EDS, was Fe (49.64%), Na (1.41%), Cl (0.58%), O (36.84%), N (3.35%), and C (8.18%). The presence of nitrogen and carbon in the EDS surface analysis result may show that the organic BIM molecule engaged in the surface film formation.

(a)   (b)

Figure 3. The appearance of bare ST37 specimens after 24 h immersion in the NaCl solution with no additives (**a**) and with BIM (**b**).

Figure 4. The result of FE-SEM test on bare sample surface after 24 h exposure to 3.5 wt.% NaCl electrolyte containing 1 mM BIM.

At the next step, different concentrations of benzimidazole (0 wt.%, 0.5 wt.%, 0.75 wt.%, and 1 wt.%) were introduced into the epoxy-polyamide coating formulation applied on ST37 substrate and cathodic disbonding test (-1.2 V vs. Ag/AgCl for 24 h) was conducted on the coated samples. The evolution of disbonded area for different samples is demonstrated in Figure 5 as bar diagrams. From the figure, the disbonded areas for the samples either with or without the organic inhibitor increased with elapsing the immersion time. Moreover, the presence of BIM had a positive impact on the cathodic disbonding resistance. To top it off, it was clearly visible that the polymeric coating resistance to cathodic disbonding is noticeably dependent on the concentration of bezimidazole. The coating containing 0.75 wt.% BIM revealed the lowest disbondment and generally the disbonded areas increased following the order 0 wt.% > 1 wt.% > 0.5 wt.% > 0.75 wt.% BIM. Inclusion of the organic inhibitor into the polymeric matrix may result in some contradictory impacts on the coating function. The disbonding front might be covered by a protective layer when the organic molecules access the substrate surface. This can improve the strength of bonding of the polymeric film to the metallic specimen and decrease the generation of hydroxyl ion as a result of restriction of electrochemical reactions on the surface. This was previously shown in the results of the solution phase study. By increasing the organic molecule content, the chance of formation of the mentioned surface film increases. On the other hand, the interaction between molecules of inhibitor and polymeric matrix may affect the curing process and final film integrity [37]. Another drawback of the addition of the organic material to the coating is uncontrollable liberation of inhibiting species [38]. The consequence of these contradictory issues was inferior and superior behavior of the epoxy coatings with one and 0.75 wt.% BIM, respectively. This was in agreement with the results of our previous works, where exceeding the optimum concentration of corrosion inhibitors adversely affected the coating performance [35,39]. In the case of organic inhibitors liberated from Mentha longifolia, the efficient concentration was 200 ppm and incorporation of 400 ppm caused coating to lose its integrity. In other words, the amount of this green inhibitor was suggested to have a negative effect on the film formation and to decrease the crosslinking density, making the film more permeable [35]. Interestingly, the disbonded surface area of epoxy-polyamide coating containing 0.75 wt.% BIM was approximately half of that of the rest after 24 h subjection to the cathodic disbonding test. The effective role of organic molecules on the cathodic disbonding resistance of polymeric coating was also shown in Izadi and coworkers research where epoxy coating was formulated by a nanocontainer doped with Nettle molecules as a green corrosion inhibitor [24].

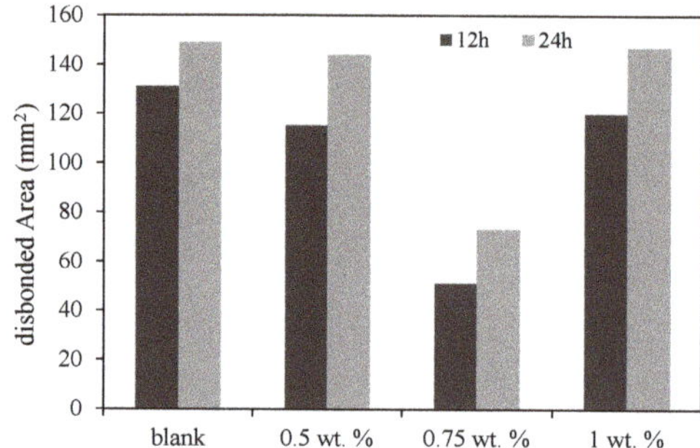

Figure 5. Evolution of delaminated area for the epoxy-polyamide coatings with different formulations on the ST37 substrate.

To provide further investigation on the effect of the concentration of BIM corrosion inhibitor on the cathodic disbonding of epoxy coating applied on ST37, AC impedance spectra were gathered. According to the literature, the cathodic delamination behavior of polymeric coatings can be analyzed through taking advantage of the EIS data [16,40,41]. For instance, Chen et al. [41] used electrochemical impedance spectroscopy for investigating the mechanism of the cathodic disbonding behavior of the three-layer polyethylene pipeline coating in chloride solution. The Nyquist plots for the ST37 panels coated with the epoxy films including different concentrations of BIM after polarization at −1.2 V vs. Ag/AgCl in 3.5 wt.% NaCl solution for 12 and 24 h are shown in Figure 6, and a typical Bode-phase and modulus diagram of the metal coated with epoxy containing 0.75 wt.% BIM is illustrated in Figure 7. Emerging as only a one time constant for all spectra caused a simple selection $R(RC)$ equivalent circuit for modeling of the data. The results of fitting with an $R(RC)$ equivalent circuit are introduced in Table 2. From the Nyquist diagrams for both exposure periods, the sample containing 0.75 wt.% BIM was characterized with the largest semicircles. While semicircles all got smaller by expanding the exposure, the sample containing 0.75 wt.% of corrosion inhibitor kept its superiority. In addition to showing the dependency of the resistance and capacitance elements on the corrosion inhibitor amount, the data presented in Table 2 revealed that the samples with 0.75 wt.% BIM possess the highest resistance and the lowest capacitance values, particularly after 24 h. In agreement with the results of a cathodic disbonding test in Figure 5, the superiority of this sample was found on the basis of the direct and inverse proportion of capacitance and resistance, respectively, to the delaminated surface area [42,43]. The substrate surface coverage with a layer composed of the organic molecules might be responsible for the behavior, as discussed before. Accordingly, the progress of delamination for all specimens, by increasing the time of subjection to the cathodic disbonding test, can also be evidenced.

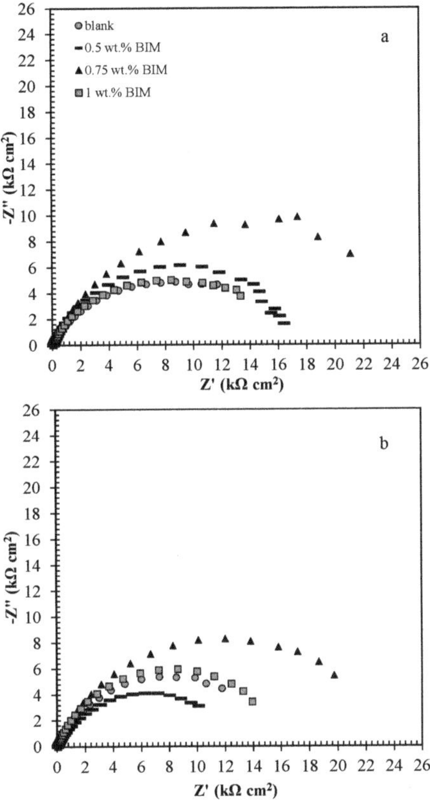

Figure 6. Nyquist plots of the ST37 panels coated with the epoxy films including different concentrations of BIM (0 wt.%, 0.5 wt.%, 0.75 wt.%, and 1 wt.%) after polarization at −1.2 V vs. Ag/AgCl in 3.5 wt.% NaCl solution for (**a**) 12 and (**b**) 24 h.

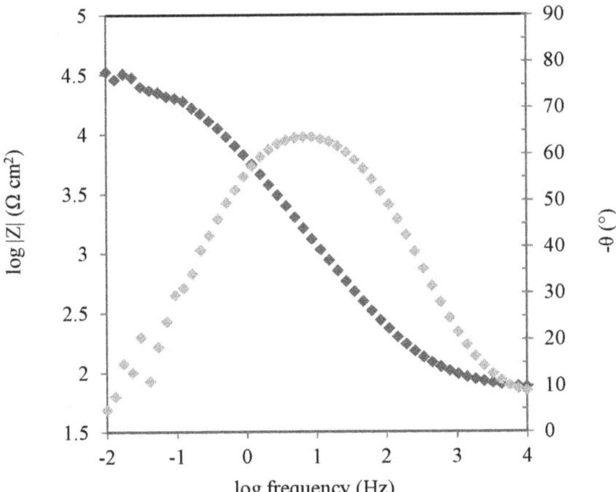

Figure 7. Typical Bode-modulus and phase diagram obtained for ST37 panel coated with the epoxy containing 0.75 wt.% of BIM after 12 h polarization at −1.2 V vs. Ag/AgCl in 3.5 wt.% NaCl solution.

Table 2. Fitting results of spectra of the ST37 panels coated with the epoxy films including different concentrations of BIM (0 wt.%, 0.5 wt.%, 0.75 wt.%, and 1 wt.%) after polarization at -1.2 V vs. Ag/AgCl in 3.5 wt.% NaCl solution for 12 and 24 h.

BIM Concentration (wt.%)	Immersion Time (h)	R (k$\Omega\cdot$cm^2)	C (μF$\cdot$cm^{-2})
0	12	15.7	39.6
	24	14.8	39.4
0.5	12	17.5	31.1
	24	12.5	43.4
0.75	12	27.8	36.2
	24	23.6	22.9
1	12	15.9	17.2
	24	14.9	46.4
Standard deviation	R	1.3%–10.1%	
	C	3%–14.4%	

To assess the effect of BIM on the adhesion strength of epoxy coating with and without 0.75 wt.% BIM (the optimum concentration previously determined in EIS and cathodic disbonding tests), pull-off test was employed. Adhesion strength values of the coatings with and without BIM before and after 24 h immersion in 3.5 wt.% NaCl are shown in Figure 8. As can be observed from the figure in the dry mode, the coating with no inhibitor revealed stronger adhesion to the ST37 substrate. This fact could be related to the increase of the coating heterogeneity as a result of the introduction of the inhibitor. The heterogeneity might lead to a decrease in the adhesion of coating film to the metal surface [10]. In contrast, the coating formulated with 0.75 wt.% BIM possessed a higher value of bonding strength after 24 of dipping in the aggressive electrolyte. As mentioned previously, in both the coating and solution phase studies, the surface layer precipitated in the presence of BIM, limiting the electrochemical reaction rate and reinforcing the interface, can increase the polymeric film adhesion. Moreover, it is also reported that the presence of the inhibitor may lead to the stabilization of the corrosion products, decreasing coating delamination [10].

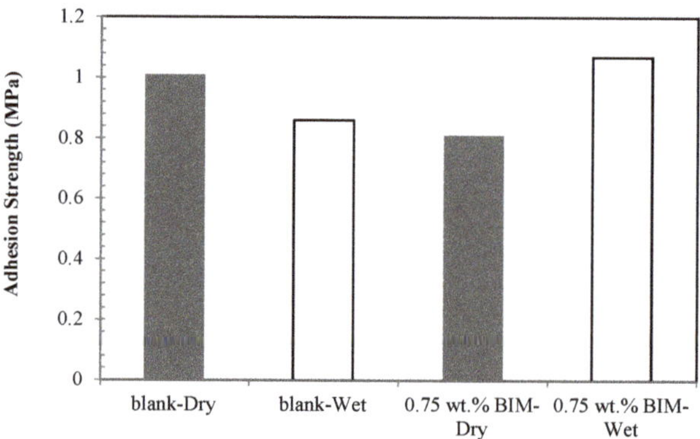

Figure 8. The impact of BIM on the bonding strength of the epoxy-polyamide coating before (dry) and after 24 h immersion in 3.5 wt.% NaCl electrolyte (wet).

The result of SEM analysis of the ST37 sample surface beneath the disbonded area of epoxy coating formulated with 0.75 wt.% BIM after 24 h polarization at -1.2 V vs. Ag/AgCl in 3.5 wt.% NaCl solution is shown in Figure 9, where deposition of a layer can be clearly observed. Moreover, detection of N (4.16 wt.%) and C (13.34 wt.%) in the EDS analysis can confirm the speculation made in the cathodic disbonding section relating to the engagement of the corrosion inhibitor in the film formation at the disbonding front. As mentioned before, the layer had an important role in the control of coating delamination through affecting the electrochemical reactions generation hydroxyl ions.

Figure 9. The result of SEM analysis of the ST37 sample surface beneath the disbonded area of epoxy coating formulated with 0.75 wt.% BIM after 24 h polarization at −1.2 V vs. Ag/AgCl in 3.5 wt.% NaCl solution.

4. Conclusions

Assessment of the influence of benzoimidazole concentration on the cathodic delamination of epoxy coating applied on ST37 substrate came down to the following conclusions:

- The solution phase study indicated that addition of benzoimidazole to chloride solution decreased the corrosion of bare ST37 sample through development of a protective layer on the surface or stabilization of the corrosion products.
- Cathodic disbonding resistance of epoxy-polyamide coating was shown to be dependent on the inhibitor content.
- Inclusion of 0.75 wt.% corrosion inhibitor to the epoxy-polyamide coating formulation decreased the cathodic disbonded surface area effectively, which is consistent with the electrochemical data.
- The wet adhesion of the polymeric coating was significantly enhanced through the addition of 0.75 wt.% benzoimidazole.

Author Contributions: Conceptualization, R.N.; Analysis, S.N.; Interpretation of Data, S.N. and N.A.; Writing–Original Draft Preparation, S.N. and N.A.; Writing–Review & Editing, R.N.; Supervision, R.N.

Funding: This research received no external funding.

Conflicts of Interest: The authors declare no conflict of interest.

References

1. Kamimura, T.; Kishikawa, H. Mechanism of cathodic disbonding of three-layer polyethylene-coated steel pipe. *Corrosion* **1998**, *54*, 979–987. [CrossRef]
2. Leidheiser, H.; Wang, W.; Lgetoft, L. The mechanism for cathodic delamination of organic coating from a metal surface. *J. Prog. Org. Coat.* **1983**, *11*, 19–40. [CrossRef]
3. Zhou, W.; Jeffers, T.E. Application temperature, cure, and film thickness affect cathodic disboniment of FBE coatings. *Mater. Perform.* **2006**, *45*, 24–28.
4. Grundmeier, G.; Stratmann, M. Adhesion and de-adhesion mechanisms at polymer/metal interfaces: Mechanistic understanding based on in situ studies of buried interfaces. *Annu. Rev. Mater.* **2005**, *35*, 571–615. [CrossRef]

5.	Nazarov, A.; Le Bozec, N.; Thierry, D. Assessment of steel corrosion and deadhesion of epoxy barrier paint by scanning Kelvin probe. *Prog. Org. Coat.* **2018**, *114*, 123–134. [CrossRef]

6.	Parhizkar, N.; Shahrabi, T.; Ramezanzadeh, B. Steel surface pre-treated by an advance and eco-friendly cerium oxidenanofilm modified by graphene oxide nanosheets; electrochemical and adhesion measurements. *J. Alloy. Compd.* **2018**, *747*, 109–123. [CrossRef]

7.	Sorensen, P.; Dam-Johansen, K.; Weinell, C.; Kiil, S. Cathodic delamination: Quantification of ionic transport rates along coating–steel interfaces. *Prog. Org. Coat.* **2010**, *68*, 70–78. [CrossRef]

8.	Martinez, S.; Zulj, L.; Kapor, F. Disbonding of underwater-cured epoxy coating caused by cathodic protection current. *Corros. Sci.* **2009**, *51*, 2253–2258. [CrossRef]

9.	Fu, A.; Cheng, Y. Characterization of corrosion of X65 pipeline steel under disbonded coating by scanning Kelvin probe. *Corros. Sci.* **2009**, *51*, 914–920. [CrossRef]

10.	Mahdavian, M.; Attar, M. The effect of benzimidazole and zinc acetylacetonate mixture on cathodic disbonding of epoxy coated mild steel. *Prog. Org. Coat.* **2009**, *66*, 137–140. [CrossRef]

11.	Leidheiser, H. Coatings. In *Corrosion Mechanisms*; Mansfeld, F., Ed.; Marcel Dekker: New York, NY, USA, 1986.

12.	Schweitzer, P.A. *Corrosion Engineering Handbook*; CRC Press: Boca Raton, FL, USA, 2006.

13.	Greenfield, D.; Scantlebury, D. The protective action of organic coatings on steel: A review. *J. Cross. Sci. Eng.* **2000**, *3*, 5.

14.	Forsgren, A. *Corrosion Control through Organic Coatings*; CRC Press: Boca Raton, FL, USA, 2006.

15.	Naderi, R.; Attar, M. The role of zinc aluminum phosphate anticorrosive pigment in protective performance and cathodic disbondment of epoxy coating. *Corros. Sci.* **2010**, *52*, 1291–1296. [CrossRef]

16.	Naderi, R.; Attar, M. Cathodic disbondment of epoxy coating with zinc aluminum polyphosphate as a modified zinc phosphate anticorrosion pigment. *Prog. Org. Coat.* **2010**, *69*, 392–395. [CrossRef]

17.	Naderi, R.; Attar, M. Effect of zinc-free phosphate-based anticorrosion pigment on the cathodic disbondment of epoxy-polyamide coating. *Prog. Org. Coat.* **2014**, *77*, 830–835. [CrossRef]

18.	Zhang, Z.; Chen, S.; Li, Y.; Li, S.; Wang, L. A study of the inhibition of iron corrosion by imidazole and its derivatives self-assembled films. *Corros. Sci.* **2009**, *51*, 291–300. [CrossRef]

19.	Tan, A.; Soutar, A. Hybrid sol–gel coatings for corrosion protection of Copper. *Thin Solid Film* **2008**, *516*, 5706–5709. [CrossRef]

20.	Yang, H. Plasma Treatment of Organic Inhibitors for Corrosion Protection of Aerospace Alloys. Master's Thesis, University of Cincinnati, Cincinnat, OH, USA, 2003.

21.	Tadokoro, K.; Shoji, H.; Sakon, T.; Jitsuhara, I.; Yamasaki, M. Metallic Sheet Having rust-Preventive Organic Coating Thereon, Process for the Production Thereof and Treating Fluid Therefore. U.S. Patent 6,254,980, 3 July 2001.

22.	Cook, R.L. Releasable Corrosion Inhibitor Compositions. U.S. Patent 6,933,046, 23 August 2005.

23.	Brooman, E.W. Modifying organic coatings to provide corrosion resistance—Part III: Organic additives and conducting polymers. *Met. Finish.* **2002**, *100*, 104–110.

24.	Izadi, M.; Shahrabi, T.; Ramezanzadeh, B. Active corrosion protection performance of an epoxy coating applied on the mild steel modified with an eco-friendly sol-gel film impregnated with green corrosion inhibitor loaded nanocontainers. *Appl. Surf. Sci.* **2018**, *440*, 491–505. [CrossRef]

25.	Mahdavian, M.; Ashhary, S. Mercapto functional azole compounds as organic corrosion inhibitors in a polyester-melamine coating. *Prog. Org. Coat.* **2010**, *68*, 259–264. [CrossRef]

26.	Ebrahimi-Mehr, M.; Shahrabi, T.; Hosseini, M. Determination of suitable corrosion inhibitor formulation for a potable water supply. *Anti-Corros. Methods Mater.* **2004**, *51*, 399–405. [CrossRef]

27.	Beiro, M.; Collazo, A.; Izquierdo, M.; Novoa, X.; Perez, C. Characterization of barrier properties of organic paints: The zinc phosphate effectiveness. *Prog. Org. Coat.* **2003**, *46*, 97–106. [CrossRef]

28.	Mousavifard, S.M.; MalekMohammadi Nouri, P.; Attar, M.M.; Ramezanzadeh, B. The effects of zinc aluminum phosphate (ZPA) and zinc aluminum polyphosphate (ZAPP) mixtures on corrosion inhibition performance of epoxy/polyamide coating. *J. Ind. Eng. Chem.* **2013**, *19*, 1031–1039.

29.	Mahdavian, M.; Naderi, R.; Peighambari, M.; Hamidpour, M.; Haddadi, S. Evaluation of cathodic disbondment of epoxy coating containing azole compounds. *J. Ind. Eng. Chem.* **2014**, *21*, 1167–1173. [CrossRef]

30. Hang, T.; Duong, N.; Truc, T.; Hoang, T.; Thanh, D.; Daopiset, S.; Boonplean, A. Effects of hydrotalcite intercalated with corrosion inhibitor on cathodic disbonding of epoxy coatings. *J. Coat. Technol. Res.* **2015**, *12*, 375–383. [CrossRef]

31. Ramezanzadeh, B.; Vakili, H.; Amini, R. The effects of addition of poly(vinyl) alcohol (PVA) as a greencorrosion inhibitor to the phosphate conversion coating on theanticorrosion and adhesion properties of the epoxy coating on thesteel substrate. *Appl. Surf. Sci.* **2015**, *327*, 174–181. [CrossRef]

32. Parhizkar, N.; Ramezanzadeh, B.; Shahrabi, T. Corrosion protection and adhesion properties of the epoxy coating applied on the steel substrate pre-treated by a sol-gel based silane coating filled with amino and isocyanate silane functionalized grapheme oxide nanosheets. *Appl. Surf. Sci.* **2018**, *439*, 45–59. [CrossRef]

33. Bentiss, F.; Jama, C.; Mernari, B.; El Attari, H.; El Kadi, L.; Lebrini, M.; Traisnel, M.; Lagrenee, M. Corrosion control of mild steel using 3,5-bis(4-methoxyphenyl)-4-amino-1,2,4-triazole in normal hydrochloric acid medium. *Corros. Sci.* **2009**, *51*, 1628–1635. [CrossRef]

34. Deflorian, F.; Rossi, S.; Fedel, M.; Motte, C. Electrochemical investigation of high-performance silane sol-gel films containing clay nanoparticles. *Prog. Org. Coat.* **2010**, *69*, 158–166. [CrossRef]

35. Nikpour, S.; Naderi, R.; Mahdavian, M. Fabrication of silane coating with improved protection performance using Mentha longifolia extract. *J. Taiwan Inst. Chem. Eng.* **2018**, *88*, 261–276. [CrossRef]

36. Bahrani, A.; Naderi, R.; Mahdavian, M. Chemical modification of talc with corrosion inhibitors to enhance the corrosion protective properties of epoxy-ester coating. *Prog. Org. Coat.* **2018**, *120*, 110–122. [CrossRef]

37. Dias, S.; Lamaka, S.; Nogueira, C.; Diamantino, T.; Ferreira, M. Sol-gel coatings modified with zeolite fillers for active corrosion protection of AA2024. *Corros. Sci.* **2012**, *62*, 153–162. [CrossRef]

38. Zheludkevich, M.; Poznyak, S.; Rodrigues, L.; Raps, D.; Hack, T.; Dick, L.; Nunes, T.; Ferreira, M. Active protection coatings with layered double hydroxide nanocontainers of corrosion inhibitor. *Corros. Sci.* **2010**, *52*, 602–611. [CrossRef]

39. Naderi, R.; Fedel, M.; Urios, T.; Poelman, M.; Olivier, M.; Deflorian, F. Optimization of silane sol-gel coatings for the protection of aluminium components of heat exchangers. *Surf. Interface Anal.* **2013**, *45*, 1457–1466. [CrossRef]

40. Ismail, I.; Harun, M. Cathodic disbonding of industrial chlorinated rubber-based primer used in rubber/metal composites: An electrochemical impedance spectroscopy analysis. *Rubber Chem. Technol.* **2016**, *89*, 712–723. [CrossRef]

41. Chen, Y.; Wang, X.; Li, Y.; Zheng, G.; Tu, X. Electrochemical impedance spectroscopy study for cathodic disbonding test technology on three layer polyethylene anticorrosive coating under full immersion and alternating dry–wet environments. *Int. J. Electrochem. Sci.* **2016**, *11*, 10884–10894. [CrossRef]

42. Hirayama, R.; Haruyama, S. Electrochemical impedance for degraded coated steel having pores. *Corrosion* **1991**, *47*, 952–957. [CrossRef]

43. Jorcin, J.; Aragon, E.; Merlatti, C.; Pebere, N. Delaminated areas beneath organic coating: A local electrochemical impedance approach. *Corros. Sci.* **2006**, *48*, 1779–1790. [CrossRef]

Article

Spectroscopic and Structural Analyses of *Opuntia Robusta* Mucilage and Its Potential as an Edible Coating

Aurea Bernardino-Nicanor [1], José Luis Montañez-Soto [2], Eloy Conde-Barajas [1], María de la Luz Xochilt Negrete-Rodríguez [1], Gerardo Teniente-Martínez [1], Enaim Aída Vargas-León [1], José Mayolo Simitrio Juárez-Goiz [1], Gerardo Acosta-García [1] and Leopoldo González-Cruz [1,*]

[1] Tecnológico Nacional de México-Celaya, Antonio García Cubas Pte #600 esq. Av. Tecnológico. Celaya, Gto. Mexico, Celaya C.P. 38010, Mexico; aurea.bernardino@itcelaya.edu.mx (A.B.-N.); eloy.conde@itcelaya.edu.mx (E.C.-B.); xochilt.negrete@itcelaya.edu.mx (M.d.l.L.X.N.-R.); gera_tm@hotmail.com (G.T.-M.); enain_32@hotmail.com (E.A.V.-L.); jmayolo@gmail.com (J.M.S.J.-G.); gerardo.acosta@itcelaya.edu.mx (G.A.-G.)

[2] Centro Interdisciplinario de Investigación para el Desarrollo Integral Regional del Instituto Politécnico Nacional, Unidad Michoacán, Justo Sierra No. 28 Jiquilpan, Mich C.P. 59510, Mexico; montasoto@yahoo.com.mx

* Correspondence: leopoldo.gonzalez@itcelaya.edu.mx; Tel.: +52-01-461-611-75-75 (ext. 5479)

Received: 28 November 2018; Accepted: 12 December 2018; Published: 15 December 2018

Abstract: Mucilage extracted from the parenchymatous and chlorenchymatous tissues of *Opuntia robusta* were obtained using water or ethanol as the extraction solvent. The changes in the different tissues by using different extraction solvents were evaluated via scanning electron microscopy (SEM) and Fourier transform infrared (FT-IR) and Raman spectroscopy; in addition, the effect of mucilage coating on the various quality characteristics of tomato (*Lycopersicum sculentum*) was evaluated. The SEM results showed that the mucilage extracted from the parenchyma had a higher aggregation level that the mucilage extracted from the chlorenchyma. The presence of three characteristic bands of pectic substances in the FT-IR spectra between 1050 and 1120 cm^{-1} indicated that the mucilage extracted from the parenchymatous tissue had a higher content of pectic compounds than the mucilage extracted from the chlorenchymatous tissue. It was also observed in the Raman spectra that the level of pectic substances in the mucilage extracted from the parenchymatous was higher than that in the mucilage extracted from the chlorenchymatous tissue. The mucilage extracted from the parenchymatous tissue was more effective as an edible coating than the mucilage extracted from the chlorenchymatous tissue. Tomatoes covered with mucilage showed significantly enhanced firmness and reduced weight loss. The uncoated tomatoes showed higher lycopene content than the coated tomatoes on the 21st day. This study showed that the *Opuntia robusta* tissue and extraction solvent influence mucilage characteristics and that *Opuntia robusta* mucilage is a promising edible coating.

Keywords: mucilage; *Opuntia robusta*; shelf life; tomatoes

---

1. Introduction

The tomato (*Lycopersicum esculentum*) is a climacteric fruit with a relatively short postharvest life during which softening and textural changes occur. In addition, many components of biochemical pathways involved in pigmentation, carbohydrate metabolism and ethylene biosynthesis are also modified [1]. Postharvest practices can have a significant effect on tomato sugar content; however, physical characteristics are the most important factor for the final consumer, and fresh tomato quality is determined by appearance, color, firmness and flavor. On the other hand during maturation, the weight loss of the tomato is increased by the moisture effect, solute movement, and water loss [2]. For this reason, diverse studies have been focusing on the control of tomato ripening through of the use of modified atmospheres, temperature and humidity control, hypobaric storage and edible coatings [3,4].

The use of edible coatings has been considered an alternative in response to the increasing demand for fresh and minimally processed tomatoes, enhancing their shelf life. The application of edible coatings in tomatoes is a promising technology since these coatings can act as a vehicle for the incorporation of fungicides and antimicrobial agents [3,5], reduce the rate of color change, and inhibit ethylene production [6]; they can also act as barriers to water loss and gas exchange, enable the controlled release of bioactive compounds, and enhance the antioxidant activity and total phenolic contents [7].

In the development of edible coatings, a broad range of edible biopolymers has been used, such as mucilage, starches, cellulose derivatives, pectin, carrageenan, chitosan/chitin, sucrose, gums, sodium alginate, proteins (animal or plant-based) and lipids [8]. The use of edible coatings has diverse advantages compared with synthetic polymers; edible coatings endow food products with a natural appearance, reduce the environmental impact, and are nontoxic, economical and easily available in the environment [9]. In addition, edible coatings offer biocompatibility and can be used as additive carriers of colorants, flavors, antioxidants or antimicrobials.

The plants of the *Opuntia* genus have high concentration of polysaccharides, principally mucilages that contain arabinose, galactose, galacturonic acid, rhamnose, and xylose [10], on the other hand, Rocchetti et al., [11] reported the presence of high content of β-glucans in the *Opuntia ficus-indica* cladodes, that are capable of forming gels in water, for this reason, the mucilaginous compounds of *Opuntia robusta* have been used in the development of foodstuffs [12]; in addition, it has been observed that the mucilage from nopal also has the ability to form edible coatings [13] and has been used to develop an edible coating that increases the shelf-life of strawberry [14].

However, few studies have been conducted using Raman, FT-IR and scanning electron microscopy (SEM) methods to compare mucilage extracted from the parenchyma and that from the chlorenchyma of *Opuntia robusta* or to examine the effect of modifying the extraction solvent. For this reason, in this study, FT-Raman, FT-IR spectroscopy and scanning electron microscopy (SEM) were used to determine the differences between mucilage extracted from the parenchyma and mucilage extracted from the chlorenchyma of *Opuntia robusta* and the effect of the extraction solvent on the structure of the mucilage. The suitability of *Opuntia robusta* mucilage as an edible coating to extend the shelf life of tomato was also evaluated.

2. Materials and Methods

2.1. Vegetative Material

Opuntia robusta is a shrubby to tree-like cactus originating in central Mexico that can attain a height of 3 m, is a source of fruits as well as young cladodes used for animal feed. Young cladodes from *Opuntia robusta* Wendl var. *robusta* were obtained from Tulancingo, Hidalgo, Mexico. The cladodes were harvested manually at 12:00 p.m. and selected according to size to normalize the state of maturity: 35 cm (large), 30 cm (width) and 4.5 cm (thickness). The samples were stored at 4 °C in a refrigerator (LG, Model GR-452SH, LG electronics, Mexico, Mexico) for up to 24 h until used.

2.2. Obtention of Mucilage

Chlorenchymatous tissue and spongy parenchymatous tissue were removed from the cladodes using a knife, and both tissues were subsequently cut into approximately 2 cm cubes, packed in PVC bags and stored in the freezer (LG Model GR-452SH) until used.

One hundred g of parenchymatous or chlorenchymatous tissue were placed in a flask (500 mL) containing a stirrer and 100 mL of ethanol or water. The mixture was stirred frequently (2 h; 50 °C) and then ground in a high-speed blender for 5 min. The mixture was allowed to stand for 1 h and filtered using Whatman (Maidstone, UK) No. 40 filter paper to obtain a fully clarified mucilaginous extract. The mucilaginous extract obtained from *Opuntia robusta* was dried (50 °C) until the complete elimination of moisture in a forced convection drying-oven (Binder, Model FD115-UL, Tuttlingen, Germany). Dried mucilage was milled into a fine powder and sieved through a size 40 mesh (425 μm). The mucilage powder was packaged in 25-g glass bottles and stored at 25 °C until used.

2.3. Scanning Electron Microscopy

Samples of *Opuntia robusta* mucilage powder were examined using a JEOL (JEOL, type EX-1200, Tokyo, Japan) scanning electron microscope fitted with a Kevex Si(Li) X-ray detector (Kevex inc, Newark, DE, USA). The analyses were performed under vacuum at an accelerating voltage of 15 kV. The samples were mounted on double-sided carbon tape and covered with approximately 10 nm of gold using a Denton sputter coater (Denton Vaccum LLC, Moorestown, NJ, USA).

2.4. FT-IR Spectroscopy

The FT-IR spectra of the *Opuntia robusta* mucilage were acquired on a Perkin Elmer FT-IR spectrophotometer (Perkin Elmer, Inc., Waltham, MA, USA) using potassium bromide (KBr) discs prepared from powered samples mixed with dry KBr. The spectra were recorded (16 scans) in transparent mode at a resolution of 4000 to 400 cm^{-1}.

2.5. Raman Spectroscopy

The Raman measurements were performed on a Perkin-Elmer (Perkin Elmer, Inc., Waltham, MA, USA) 2000R NIR FT-Raman Spectrometer equipped with a Nd:YAG laser emitting at a wavelength of 1064 nm and an InGaAs detector. For these analyses, the 180° backscattering refractive geometry was used. The spectrometer was managed using Perkin-Elmer Spectrum software (Version 3.02.00 [2000]). The spectral data for rice bean starch were obtained at a wavenumber resolution of 4 cm^{-1} and at a nominal laser power of 500 mW. For each spectrum, 20 scans were accumulated to ensure an acceptable signal-to-noise ratio. All Raman spectra were collected at room temperature.

2.6. Edible Coating Preparation

Six different treatments were prepared, which differed in the solvent (water or ethanol) and cladode tissue (parenchyma or chlorenchyma) used, for the reconstitution of the mucilage powder (Table 1). For the preparation of the edible coatings, 50 g of the mucilage powder were placed in a flask (500 mL), and then water or ethanol was added to obtain a solution of mucilage with 12% soluble solids; this procedure was carried out for each cladode tissue.

Table 1. Formulation of mucilage-based edible coatings.

Edible Coating Sample	% Soluble Solids	Solvent	Cladode Tissue
C1	12	Water	parenchyma
C2	12	Ethanol	parenchyma
C3	12	Mix *	parenchyma
C4	12	Water	chlorenchyma
C5	12	Ethanol	chlorenchyma
C6	12	Mix *	chlorenchyma

* Water:Ethanol 50:50 vol.% (relation obtained in previous experiments; data not showed).

Edible Coating Application

Prior to the application of edible coatings, the tomatoes were washed with distilled water. The coatings were applied uniformly by brushing on various tomatoes (26 pieces), three times. The tomatoes were stored at 20 °C. A control lot of tomatoes (26 pieces) was used (without edible coating).

2.7. Weight Loss During Postharvest Storage

Weight loss during postharvest storage was determined by subtracting the sample weight from their previous recorded weight and was presented as % of weight loss compared to the initial weight. Weight loss was calculated using the Equation (1).

$$\text{weight loss (\%)} = \frac{\text{Initial weight (g)} - \text{final weight (g)}}{\text{Initial weight (g)}} \times 100 \tag{1}$$

2.8. Texture Measurement

A penetration test was performed on the skin of the whole fruit using a TA.XT2i texture analyzer (Stable Micro Systems, Surrey, UK) with a 2 mm diameter cylindrical probe. The samples were penetrated to a depth of 6 mm. The speed of the probe was 1.0 mm·s^{-1} during the pretest and penetration. The tomatoes were placed in a way that the probe penetrated their equatorial zone.

2.9. Determination of Lycopene

Fifty g of tomato were ground to a homogeneous puree using a hand-held mixer (Braun Inc., Lynnfield, MA, USA). The puree was mixed with 50 mL of a hexane-acetone-ethanol mixture (50:25:25) and placed in an orbital shaker (Eberbach Corp., Ann Arbor, MI, USA) (200 rpm) for 15 min. Thereafter, 3 mL of distilled water were added, and the sample was shaken for another 15 min and then vacuum-filtered. The combined filtrates were mixed with 50 mL of hexane in a separatory funnel, and 200 mL of double-distilled water were then added; the mixture was allowed to stand for 15 min to enable phase separation. The water–acetone–ethanol phase was discarded; the hexane phase was collected into an amber screw-capped vial and concentrated to dryness with high purity nitrogen (99.99%).

Lycopene was quantified using an external standard, and the absorbance was taken at 503 nm using a spectrophotometer (Thermo Fisher Scientific Inc., Waltham, MA, USA); hexane was used as solvent blank. The results were expressed as mg·kg^{-1} of sampler.

2.10. Statistical Analysis

Quantitative data are expressed as the mean ± standard deviation, and the results were statically analyzed with ANOVA and Tukey's test at the 95% confidence level. SAS software (Statistical Analysis System V. 8.0) was used for the data analysis, and all experimental determinations were performed in triplicate.

3. Results and Discussion

3.1. Characterization of Mucilage

3.1.1. Morphology of *Opuntia* Mucilage

Scanning electron microphotographs (SEM) of the mucilage obtained from *Opuntia robusta* are shown in Figure 1. A higher aggregation level of small particles was observed in the mucilage extracted from the parenchyma (Figure 1b) than in the mucilage extracted from the chlorenchyma (Figure 1a). The particles of the mucilage extracted from the chlorenchyma of *Opuntia robusta* are mostly seen as aggregates of irregular shapes and dimensions and are fibrous in nature. Superimposed fibers have been observed in chia mucilage and increase the complexity of the aggregates [15]; apparently, a similar behavior can be observed in the mucilage extracted from the parenchyma of *Opuntia robusta*. On the other hand, the high aggregation observed in the mucilage obtained from *Opuntia robusta* is in concordance with those reported by du Toit et al. [16], indicating that *Opuntia robusta* has a higher fiber content that other *Opuntia* species.

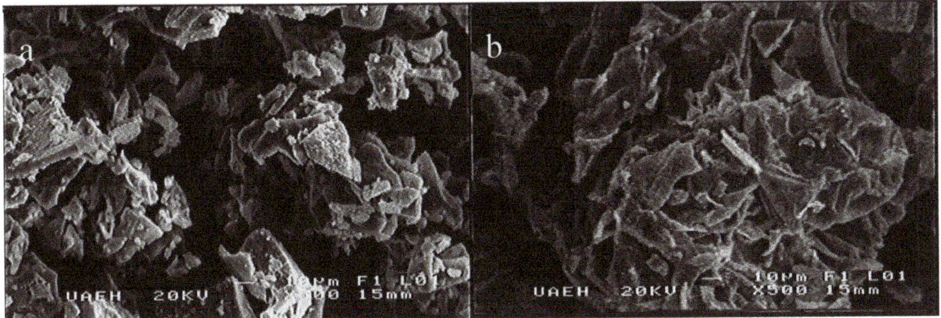

Figure 1. SEM micrograph of *Opuntia robusta* mucilage: (**a**) extracted from the chlorenchymatous tissue; (**b**) extracted from the parenchymatous tissue.

3.1.2. FT-IR Analysis of Structural Changes in Mucilage Due to the Extraction Process and the Source

The FT-IR spectra of the mucilage extracted using either water or ethanol are shown in Figure 2. It can be observed that the spectra of the mucilage present a broad, strong signal at approximately 3400 cm^{-1} due to the stretching vibration of O–H bonds; it can be observed that in both parenchyma and chlorenchyma tissues, this band was smoother in the spectrum of mucilage extracted with ethanol (Figure 2a,c) than in the spectrum of mucilage extracted with water (Figure 2b,d). Apparently, this difference is due to the formation of more hydrogen bonds in the mucilage by the interaction with ethanol. The best defined band was observed at approximately 2920 cm^{-1} for the ethanol-extracted mucilages of both tissues (parenchyma and chlorenchyma), and this band was assigned to the stretching vibration of C–H bonds from pyranose groups [17] or C–H group of the methyl ester of galacturonic acid [18].

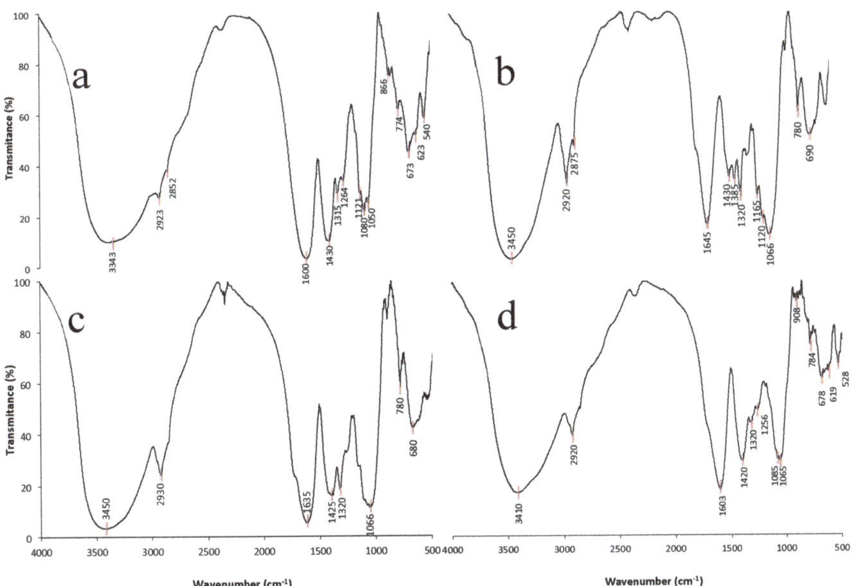

Figure 2. FT-IR patterns of *Opuntia robusta* mucilage: (**a**) extracted from the parenchyma with ethanol; (**b**) extracted from the parenchyma with water; (**c**) extracted from the chlorenchyma with ethanol; (**d**) extracted from the chlorenchyma with water.

However, some authors have indicated that this band can be attributed to proteins [19,20]; in addition, the water-extracted mucilages of both tissues show that the best defined band near 2870 cm^{-1} overlaps with the stretching vibration of the C–H bond from glucose units and with the –CH$_2$ symmetric stretching vibration of proteins derived from carboxylic acids, indicating the presence of protein residues on the mucilage [19,21].

The broad bands observed between 1600 and 1640 cm^{-1} correspond to the scissor vibrations of –OH due to bound water [17] or the stretching vibration of the carboxyl group –COOH [19,22,23]. In mucilage extracted with water or ethanol, the presence of a free carboxyl group was observed, represented by the band at approximately 1430 cm^{-1} [22]. The best defined band for the mucilage extracted with ethanol was observed at approximately 1320 cm^{-1}, indicating the presence of *o*–acetyl groups; other authors have attributed this band to the glycosidic linkage [24]. The three characteristic bands of pectic fractions [22] were observed only in the mucilage extracted from the parenchyma (1120, 1080, and 1050 cm^{-1}); in the mucilage extracted from the chlorenchyma, an overlap was observed at approximately 1085 and 1065 cm^{-1}.

3.1.3. Raman Spectroscopy Analysis of Structural Changes in the Mucilage Due to the Extraction Process and the Source

The Raman spectra of the mucilage samples from *Opuntia robusta* are presented in Figure 3. The principal differences observed in the spectra of mucilage extracted with ethanol or water occurred in the principal bands associated with mucilage; on the other hand, differences in relative intensity were also observed between the patterns of Raman spectra of the mucilage extracted with ethanol (Figure 3b,d) and those of the mucilage extracted with water (Figure 3a,c) and between the spectra of the mucilage extracted from the chlorenchyma and that extracted from the parenchyma. In the mucilage extracted from the parenchyma, a well-defined band at 2930 cm^{-1} was observed in the sample in comparison with the mucilage extracted from the chlorenchyma, which presented a change in position; in addition, overlapping bands were observed (Figure 3a).

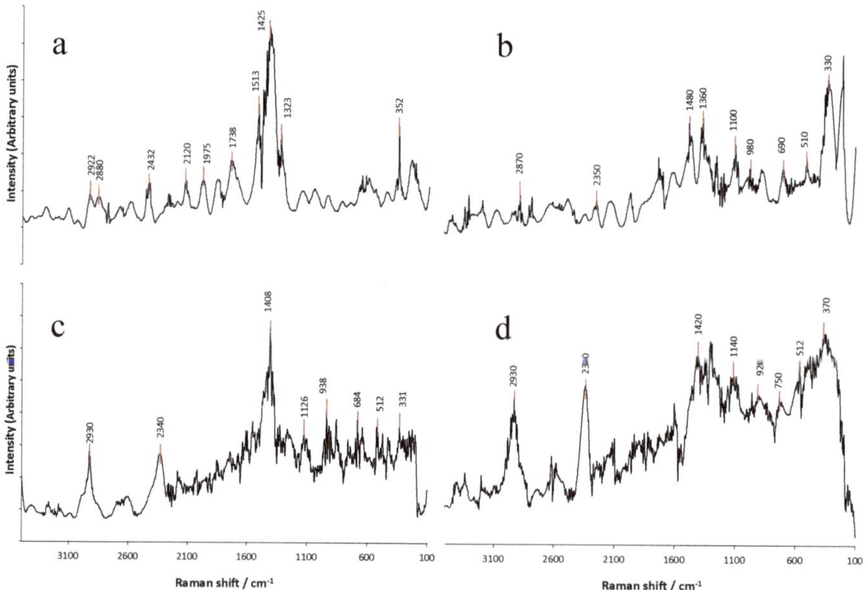

Figure 3. FT-Raman patterns of *Opuntia robusta* mucilage: (**a**) extracted from the chlorenchyma with water; (**b**) extracted from the chlorenchyma with ethanol; (**c**) extracted from the parenchyma with water; (**d**) extracted from the parenchyma with ethanol.

Some authors have indicated that the 2930 cm^{-1} band is related to the C–H stretching band of pectin, while the overlapping band at 2880 cm^{-1} is related to hemicellulose [25]; however, other authors have reported that this band is present in the mucilage of chia seeds [15].

The vibrational bands between 1400 and 1470 cm^{-1} are related to the elongation of the CH$_2$ groups of xylose, galactose and arabinose, and the CH$_3$ group of rhamnose; all of these bands were present for the *Opuntia* mucilage [10], and a higher intensity of these bands was observed in the water-extracted mucilage from the chlorenchyma than in the mucilage extracted from the parenchyma or extracted with ethanol; the intensity depends on the degree of crystallinity.

Additionally, the bands observed at 1350, 1270 and 1076 cm^{-1} are assigned as C–O–H related modes and probably correspond to residual galacturonic acid; however, some authors have indicated that the band at approximately 1330 cm^{-1} is related to the CH$_2$ vibration of α-glucose and that the band at approximately 370 cm^{-1} is assigned to the skeletal vibrational mode of δ_s(C–C) [15].

3.2. Mucilage Used as Edible Coating

3.2.1. Weight Loss of Tomatoes

The tomatoes coated with parenchyma as an edible coating tended to have smaller weight loss than the control tomatoes (Figure 4). No significant difference in weight loss was observed in the three lots of tomatoes covered with parenchyma (C1, C2 and C3); however, the weight loss of the coated tomatoes was 7 wt.% lower than that of the control tomatoes (Figure 4). On the other hand, for the tomatoes coated with chlorenchyma, the differences between the three mucilage coatings tested (C4, C5 and C6) and the control (Figure 4) were significant. The tomatoes coated with chlorenchyma as an edible coating showed higher weight loss than the tomatoes coated with parenchyma; however, smaller weight loss was observed compared to that of the uncoated tomatoes (Figure 4).

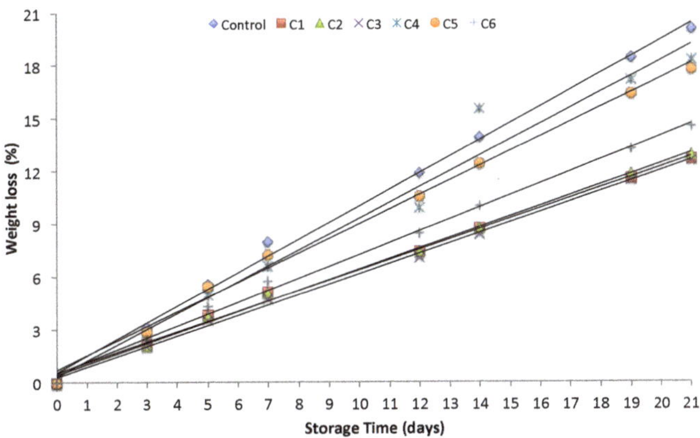

Figure 4. Effect of mucilage coating on weight loss of tomatoes during the 21-day storage period (20 ± 2 °C): (C1) edible coating sample prepared with parenchyma mucilage + water; (C2) edible coating sample prepared with parenchyma mucilage + ethanol; (C3) edible coating sample prepared with parenchyma mucilage + water:ethanol (50:50 vol.%); (C4) edible coating sample prepared with chlorenchyma mucilage + water; (C5) edible coating sample prepared with chlorenchyma mucilage + ethanol; (C6) edible coating sample prepared with chlorenchyma mucilage + water:ethanol (50:50 vol.%).

The differences between the weight loss of the tomatoes coated with mucilage extracted from the parenchyma and that of tomatoes coated with mucilage extracted from the chlorenchyma could be due to the different compositions of the mucilage sources; however, it can be inferred that either mucilage source (parenchyma or chlorenchyma) formed a semipermeable layer that reduced respiration and transpiration on the tomato surface. It has been observed that coatings confer a physical barrier against O_2, CO_2, moisture and solute movement, consequently reducing water loss and weight loss [26–28].

Our results are in agreement with the finding of Del-Valle et al. [14], where the mucilage coating from *Opuntia ficus-indica* was effective in extending the shelf-life and other postharvest parameters of quality of strawberry; in other studies, the weight loss of breba fig was strongly influenced by *Opuntia ficus-indica* mucilage coating [29].

3.2.2. Effect on the Firmness of Tomatoes

Tomato firmness decreased significantly during storage in all coated and uncoated tomatoes (Figure 5); however, the coated tomatoes showed resistance against the loss of firmness and maintained texture during the storage period, particularly those tomatoes coated by mucilage extracted from the nopal parenchyma (Figure 5). The reduction in firmness of the uncoated sample was 62% compared to 47% for the tomatoes coated with mucilage extracted from the parenchyma and 60% for the tomatoes coated with mucilage extracted from the chlorenchyma (Figure 5). Similar firmness reduction trends were observed by Allegra [29], who showed that the loss of firmness in tomatoes coated with *Opuntia ficus indica* mucilage was significantly delayed during the storage period.

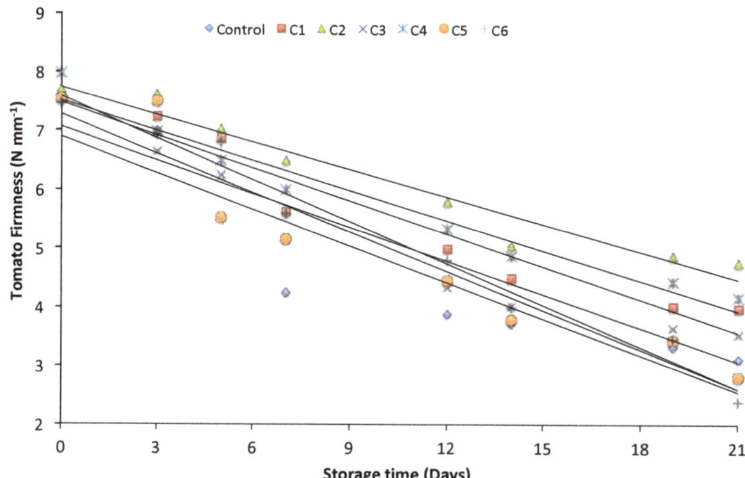

Figure 5. Effect of mucilage coating on the firmness of tomatoes during the 21-day storage period (20 ± 2 °C): (C1) edible coating sample prepared with parenchyma mucilage + water; (C2) edible coating sample prepared with parenchyma mucilage + ethanol; (C3) edible coating sample prepared with parenchyma mucilage + water:ethanol (50:50 vol.%); (C4) edible coating sample prepared with chlorenchyma mucilage + water; (C5) edible coating sample prepared with chlorenchyma mucilage + ethanol; (C6) edible coating sample prepared with chlorenchyma mucilage + water:ethanol (50:50 vol.%).

Surface coatings have been found to retain higher firmness, and the positive effect is attributed to the restriction in metabolic activities associated with cell wall-degrading enzymes or to the calcium content in species of the *Opuntia* genus [12,30], since it has been observed that calcium treatment maintains the firmness of peach, fig and strawberry [31–33].

3.2.3. Effect on the Lycopene Content of Tomatoes

The lycopene contents of the tomatoes coated and uncoated with mucilage during the red stage on the 21st day are shown in Table 2. In general, the results are in concordance with those of other studies on lycopene content in tomatoes that indicated that the lycopene content of stored tomatoes is in the range of 78.6 and 324 mg·kg^{-1} of tomato on a fresh weight basis and between 1710 and 4550 mg·kg^{-1} of tomato on a dry weight basis [34,35], however the extraction method was determinant to the extractable amount of lycopene and purity, reaching in some cases around purity level over 95% [36]. The lycopene content of the uncoated tomatoes on the 21st day of storage was 20% and 6% higher than those of the tomatoes coated with mucilage extracted from the parenchyma and the chlorenchyma, respectively.

Table 2. Effect of coating on the lycopene content of tomato on the 21st day of storage.

Sample	Lycopene Content (mg·100 g)	
	Fresh Weight Basis	Dry Weight Basis
Control	168 ± 12	3685 ± 250
C1	132 ± 9.0	2714 ± 184
C2	137 ± 11	2854 ± 218
C3	143 ± 13	3069 ± 284
C4	163 ± 14	3456 ± 303
C5	160 ± 18	3385 ± 299
C6	152 ± 12	3096 ± 263

Our results are in agreement with those of previous reports suggesting that the formation of lycopene depends on the rate of respiration during storage [37,38]; apparently, the mucilage provided a thick barrier against ethylene production and gas exchange between the inner and outer environments and therefore delayed the ripening of the fruit during storage [2].

4. Conclusions

Differences in morphology between the mucilage extracted from the parenchymatous tissue and the mucilage extracted from the chlorenchymatous tissue were detected. The FT-IR and Raman spectroscopy data showed that the mucilage extracted from the parenchymatous tissue had an apparently higher level of pectic substances than the mucilage extracted from the chlorenchymatous tissue; in addition, changes in the FT-IR and Raman spectra patterns were observed due to the extraction solvent. The scanning electron microscopy (SEM) showed higher aggregation level in the mucilage extracted from the parenchyma than in the mucilage extracted from chlorenchyma. Tomatoes covered with mucilage showed significantly enhanced firmness and reduced weight loss, while the uncoated tomatoes showed higher lycopene content than the coated tomatoes on the 21st day. Finally, this study showed that *Opuntia robusta* mucilage is a promising edible coating and that tissue and extraction solvent influence mucilage characteristics.

Author Contributions: Conceptualization, A.B.-N. and G.-C.L.; Formal Analysis, E.C.-B. and M.d.l.L.X.N.-R.; Funding Acquisition, A.B.-N. and G.-C.L.; Investigation, J.L.M.-S. and G.A.-G.; Methodology, E.A.V.-L.; Resources, G.T.-M. and J.M.S.J.-G.; Validation, G.T.-M. and J.M.S.J.-G.; Writing–Original Draft, A.B.-N. and G.-C.L.; Writing–Review & Editing, G.A.-G.; Supervision, A.B.-N. and G.-C.L.

Funding: This research was funded by National Technologic of Mexico (TecNM, No. 6405.18-P).

Acknowledgments: The authors would like to acknowledge the Autonomous University of Hidalgo State for technical services during this study.

Conflicts of Interest: The authors declare no conflict of interest.

References

1. Alexander, L.; Grierson, D. Ethylene biosynthesis and action in tomato: A model for climacteric fruit ripening. *J. Exp. Bot.* **2002**, *53*, 2039–2055. [CrossRef] [PubMed]
2. Ali, A.; Maqbool, M.; Ramachandran, S.; Alderson, P.G. Gum arabic as a novel edible coating for enhancing shelf-life and improving postharvest quality of tomato (*Solanum lycopersicum* L.) fruit. *Postharvest Biol. Technol.* **2010**, *58*, 42–47. [CrossRef]

3. Rodríguez-Garcia, I.; Cruz-Valenzuela, M.R.; Silva-Espinoza, B.A.; Gonzalez-Aguilar, G.A.; Moctezuma, E.; Gutierrez-Pacheco, M.M.; Tapia-Rodríguz, M.R.; Ortega-Ramírez, L.A.; Ayala-Zavala, J.F. Oregano (*Lippia graveolens*) essential oil added within pectin edible coatings prevents fungal decay and increases the antioxidant capacity of treated tomatoes. *J. Sci. Food Agric.* **2016**, *96*, 3772–3778. [CrossRef] [PubMed]

4. Ahmed, M.F.; Islam, M.Z.; Sarker, M.S.H.; Hasan, S.K.; Mizan, R. Development and performance study of controlled atmosphere for fresh tomato. *World J. Eng. Tech.* **2016**, *4*, 168–175. [CrossRef]

5. Vodnar, D.C.; Pop, O.L.; Dulf, F.V.; Socaciu, C. Antimicrobial efficiency of edible films in food industry. *Notulae Botanicae Horti Agrobotanici Cluj-Napoca* **2015**, *43*, 302–312. [CrossRef]

6. Zapata, P.J.; Guillén, F.; Martínez-Romero, D.; Castillo, S.; Valero, D.; Serrano, M. Use of alginate or zein as edible coatings to delay postharvest ripening process and to maintain tomato (*Solanum lycopersicon* Mill) quality. *J. Sci. Food Agric.* **2008**, *88*, 1287–1293. [CrossRef]

7. Limchoowong, N.; Sricharoen, P.; Techawongstien, S.; Chanthai, S. An iodine supplementation of tomato fruits coated with an edible film of the iodide-doped chitosan. *Food Chem.* **2016**, *200*, 223–229. [CrossRef]

8. Saha, A.; Tyagi, S.; Gupta, R.K.; Tyagi, Y.K. Natural gums of plant origin as edible coatings for food industry applications. *Crit. Rev. Biotechnol.* **2017**, *37*, 959–973. [CrossRef]

9. Sánchez-Ortega, I.; García-Almendárez, B.E.; Santos-López, E.M.; Reyes-González, L.R.; Regalado, C. Characterization and antimicrobial effect of starch-based edible coating suspensions. *Food Hydrocoll.* **2016**, *52*, 906–913. [CrossRef]

10. Trachtenberg, S.; Mayer, A.M. Composition and properties of *Opuntia ficus-indica* mucilage. *Phytochemistry* **1981**, *20*, 2665–2668. [CrossRef]

11. Rocchetti, G.; Pellizzoni, M.; Montesano, D.; Lucini, L. Italian *Opuntia ficus-indica* cladodes as rich source of bioactive compounds with health-promoting properties. *Foods* **2018**, *7*, 24. [CrossRef] [PubMed]

12. Bernardino-Nicanor, A.; Hinojosa-Hernández, E.N.; Juárez-Goiz, J.M.S.; Montañez-Soto, J.L.; Ramírez-Ortiz, M.E.; González-Cruz, L. Quality of *Opuntia robusta* and its use in development of mayonnaise-like product. *J. Food Sci. Technol.* **2015**, *52*, 343–350. [CrossRef] [PubMed]

13. Espino-Díaz, M.; De Jesús Ornelas-Paz, J.; Martínez-Téllez, M.A.; Santillán, C.; Barbosa-Cánovas, G.V.; Zamudio-Flores, P.B.; Olivas, G.I. Development and characterization of edible films based on mucilage of *Opuntia ficus-Indica* (L.). *J. Food Sci.* **2010**, *75*, E347–E352. [CrossRef] [PubMed]

14. Del-Valle, V.; Hernández-Muñoz, P.; Guarda, A.; Galotto, M.J. Development of a cactus-mucilage edible coating (*Opuntia ficus indica*) and its application to extend strawberry (*Fragaria ananassa*) shelf-life. *Food Chem.* **2005**, *91*, 751–756. [CrossRef]

15. de la Paz Salgado-Cruz, M.; Calderón-Domínguez, G.; Chanona-Pérez, J.; Farrera-Rebollo, R.R.; Méndez-Méndez, J.V.; Díaz-Ramírez, M. Chia (*Salvia hispanica* L.) seed mucilage release characterisation. A microstructural and image analysis study. *Ind. Crop. Prod.* **2013**, *51*, 453–462. [CrossRef]

16. du Toit, A.; de Wit, M.; Hugo, A. Cultivar and harvest month influence the nutrient content of *Opuntia* spp. cactus pear cladode mucilage extracts. *Molecules* **2018**, *23*, 916–928. [CrossRef] [PubMed]

17. Ishurd, O.; Zgheel, F.; Elghazoun, M.; Elmabruk, M.; Kermagi, A.; Kennedy, J.F.; Knill, C.J. A novel $(1 \rightarrow 4)$-α-d-glucan isolated from the fruits of *Opuntia ficus indica* (L.) Miller. *Carbohydr. Polym.* **2010**, *82*, 848–853. [CrossRef]

18. Bayar, N.; Kriaa, M.; Kammoun, R. Extraction and characterization of three polysaccharides extracted from *Opuntia ficus indica* cladodes. *Int. J. Biol. Macromol.* **2016**, *92*, 441–450. [CrossRef]

19. Contreras-Padilla, M.; Rodríguez-García, M.E.; Gutiérrez-Cortez, E.; del Carmen Valderrama-Bravo, M.; Rojas-Molina, J.I.; Rivera-Muñoz, E.M. Physicochemical and rheological characterization of *Opuntia ficus* mucilage at three different maturity stages of cladode. *Eur. Polym. J.* **2016**, *78*, 226–234. [CrossRef]

20. Lian, X.; Wang, C.; Zhang, K.; Li, L. The retrogradation properties of glutinous rice and buckwheat starches as observed with FT-IR, ^{13}C NMR and DSC. *Int. J. Biol. Macromol.* **2014**, *64*, 288–293. [CrossRef]

21. Díaz, K.M.; Reyes, T.F.; Cabrera, L.P.; Sánchez, M.D.; García, M.A.; de Posada Piñán, E. Characterization of laser-treated *Opuntia* using FT-IR spectroscopy and thermal analysis. *Appl. Phys. A* **2013**, *112*, 221–224. [CrossRef]

22. Lefsih, K.; Delattre, C.; Pierre, G.; Michaud, P.; Aminabhavi, T.M.; Dahmoune, F.; Madani, K. Extraction, characterization and gelling behavior enhancement of pectins from the cladodes of *Opuntia ficus indica*. *Int. J. Biol. Macromol.* **2016**, *82*, 645–652. [CrossRef] [PubMed]

23. Jadhav, M.V.; Mahajan, Y.S. Assessment of feasibility of natural coagulants in turbidity removal and modeling of coagulation process. *Desalin. Water Treat.* **2014**, *52*, 5812–5821. [CrossRef]

24. Kalegowda, P.; Chauhan, A.S.; Urs, S.M.N. *Opuntia dillenii* (Ker-Gawl) Haw cladode mucilage: Physico-chemical, rheological and functional behavior. *Carbohydr. Polym.* **2017**, *157*, 1057–1064. [CrossRef] [PubMed]

25. Chylińska, M.; Szymańska-Chargot, M.; Zdunek, A. FT-IR and FT-Raman characterization of non-cellulosic polysaccharides fractions isolated from plant cell wall. *Carbohydr. Polym.* **2016**, *154*, 48–54. [CrossRef] [PubMed]

26. Ruelas-Chacon, X.; Contreras-Esquivel, J.C.; Montañez, J.; Aguilera-Carbo, A.F.; Reyes-Vega, M.L.; Peralta-Rodriguez, R.D.; Sanchéz-Brambila, G. Guar gum as an edible coating for enhancing shelf-life and improving postharvest quality of roma tomato (*Solanum lycopersicum* L.). *J. Food Qual.* **2017**, *2017*, 8608304. [CrossRef]

27. Cipolatti, E.P.; Kupski, L.; Rocha, M.D.; Oliveira, M.D.S.; Buffon, J.G.; Furlong, E.B. Application of protein-phenolic based coating on tomatoes (*Lycopersicum esculentum*). *Food Sci. Technol.* **2012**, *32*, 594–598. [CrossRef]

28. Clifford, S.C.; Arndt, S.K.; Popp, M.; Jones, H.G. Mucilages and polysaccharides in *Ziziphus* species (Rhamnaceae): Localization, composition and physiological roles during drought-stress. *J. Exp. Bot.* **2002**, *53*, 131–138. [CrossRef]

29. Allegra, A.; Sortino, G.; Inglese, P.; Settanni, L.; Todaro, A.; Gallotta, A. The effectiveness of *Opuntia ficus-indica* mucilage edible coating on post-harvest maintenance of 'Dottato'fig (*Ficus carica* L.) fruit. *Food Packag. Shelf Life* **2017**, *12*, 135–141. [CrossRef]

30. Sepúlveda, E.; Sáenz, C.; Aliaga, E.; Aceituno, C. Extraction and characterization of mucilage in *Opuntia* spp. *J. Arid. Environ.* **2007**, *68*, 534–545. [CrossRef]

31. Irfan, P.K.; Vanjakshi, V.; Prakash, M.K.; Ravi, R.; Kudachikar, V.B. Calcium chloride extends the keeping quality of fig fruit (*Ficus carica* L.) during storage and shelf-life. *Postharvest Biol. Technol.* **2013**, *82*, 70–75. [CrossRef]

32. Manganaris, G.A.; Vasilakakis, M.; Diamantidis, G.; Mignani, I. The effect of postharvest calcium application on tissue calcium concentration, quality attributes, incidence of flesh browning and cell wall physicochemical aspects of peach fruits. *Food Chem.* **2007**, *100*, 1385–1392. [CrossRef]

33. García, J.M.; Herrera, S.; Morilla, A. Effects of postharvest dips in calcium chloride on strawberry. *J. Agric. Food. Chem.* **1996**, *44*, 30–33. [CrossRef]

34. George, B.; Kaur, C.; Khurdiya, D.S.; Kapoor, H.C. Antioxidants in tomato (*Lycopersium esculentum*) as a function of genotype. *Food Chem.* **2004**, *84*, 45–51. [CrossRef]

35. Fattore, M.; Montesano, D.; Pagano, E.; Teta, R.; Borrelli, F.; Mangoni, A.; Seccia, S.; Albrizio, S. Carotenoid and flavonoid profile and antioxidant activity in "Pomodorino Vesuviano" tomatoes. *J. Food Compos. Anal.* **2016**, *53*, 61–68. [CrossRef]

36. Montesano, D.; Fallarino, F.; Cossignani, L.; Bosi, A.; Simonetti, M.S.; Puccetti, P.; Damiani, P. Innovative extraction procedure for obtaining high pure lycopene from tomato. *Eur. Food Res. Technol.* **2008**, *226*, 327. [CrossRef]

37. Abebe, Z.; Tola, Y.B.; Mohammed, A. Effects of edible coating materials and stages of maturity at harvest on storage life and quality of tomato (*Lycopersicon Esculentum* Mill.) fruits. *Afr. J. Agric. Res.* **2017**, *12*, 550–565. [CrossRef]

38. Javanmardi, J.; Kubota, C. Variation of lycopene, antioxidant activity, total soluble solids and weight loss of tomato during postharvest storage. *Postharvest Biol. Technol.* **2006**, *41*, 151–155. [CrossRef]

Article

Self-Stratification of Ternary Systems Including a Flame Retardant Liquid Additive

Agnes Beaugendre [1], Stephanie Degoutin [1], Severine Bellayer [1], Christel Pierlot [2], Sophie Duquesne [1], Mathilde Casetta [1] and Maude Jimenez [1,*]

[1] University of Lille, ENSCL, CNRS, UMR 8207, UMET, Unité Matériaux et Transformations, F 59000 Lille, France; agnes.beaugendre@gmail.com (A.B.); stephanie.degoutin@univ-lille.fr (S.D.); severine.bellayer@univ-lille.fr (S.B.); sophie.duquesne@univ-lille.fr (S.D.); mathilde.casetta@univ-lille.fr (M.C.)
[2] University of Lille, CNRS, UMR 8181, EA CMF-4478/Unité de Catalyse et de Chimie du Solide, F 59000 Lille, France; christel.pierlot@univ-lille.fr
*   Correspondence: maude.jimenez@univ-lille.fr

Received: 3 November 2018; Accepted: 5 December 2018; Published: 6 December 2018

Abstract: Particular coating compositions based on incompatible polymer blends can produce coatings having complex layered structures after film formation. The most traditional approaches to their structural modification are the introduction of additives (extenders, inorganic pigments, surface active agents, etc.). As minor additives, some are capable of substantially accelerating the phase separation process with a moderate or negligible influence on the composition equilibrium of solutions. In contrast, in order to be effective, some have to be introduced in significant amounts, thereby substantially changing the resulting distribution of components through the film. Up to now, most of the liquid additives that have been tested destabilized the solutions while impacting the layering process. In this work, two phosphorus based liquid fillers have been introduced (at 2.5 and 5 wt.%) in a partially incompatible polymer blend based on a silicone resin and a curable epoxy resin to fire retard a polycarbonate matrix. Self-stratification was evidenced by microscopic and chemical analyses, flammability by Limiting Oxygen Index (LOI) and UL-94 tests, fire performances by Mass Loss Calorimetry and thermal stability by using a tubular furnace and ThermoGravimetric Analysis. The ternary compositions including 5 wt.% of additives exhibit the best stratification and excellent adhesion onto polycarbonate. Improvements of the fire resistant properties were observed (+7% for the LOI compared to the virgin matrix) when a 200 μm wet thick coating was applied, due to reduced flame propagation and dripping.

Keywords: self-stratifying coating; phase separation; incompatibility; solution equilibrium; flame retardancy; epoxy resin; silicon resin; liquid phosphate

---

1. Introduction

In the field of coatings science, significant improvements have been noted in properties such as chemical and fire resistance, antifouling, corrosion protection, flexibility and impermeability by using new approaches, such as the application of inorganic/organic hybrid coating systems, self-stratifying coatings, and so on [1–5]. These properties depend on the polymer selected as the bare resin of the formulation. Coatings in the industry are complex multilayer systems that must be designed to possess specific properties: the air-coating interface has to protect the underlying coatings against abrasion, chemical/solvents and UV-light, and to provide excellent impact resistance. The intermediate coating provides color, special effects and properties such as flame retardancy. Finally, the primer promotes adhesion to the substrate and protects against molecular diffusion (and corrosion in the case of a steel substrate). Each coating has a specific purpose; however, the successive application of multiple coatings is time consuming and leads to increased expenses as well as potential drawbacks such as, for example, poor interfacial adhesion.

During paint application and film formation, a polymer solution simultaneously undergoes mass and heat transfers, surface phenomena and sometimes also chemical reactions, especially in the presence of a two-component system. Phase separation of binders, from an initially constituted homophase solution, can occur under certain conditions leading to self-stratification, that is, to the formation of non-homogeneous in-layer coatings during after film formation (Figure 1). In addition, in the presence of reactive systems, further reactions like oxidation or crosslinking (e.g., two-component epoxies and urethanes) are required to generate a cross-linked film of higher molecular weight. During the reaction, the viscosity, density, and the modulus of elasticity increase due to an increase in the molar mass and crosslinking. Besides interfering with the self-stratifying process, it allows the formation of a harder primer, less flexible and much less susceptible to damage from chemical, weather or UV rays [6,7].

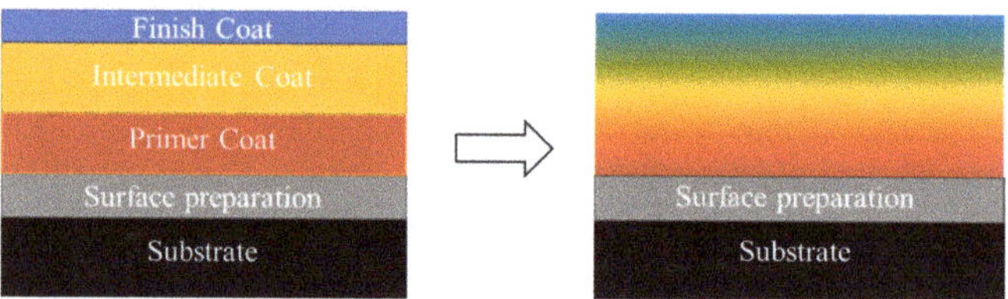

Figure 1. Self-layering concept.

Besides polymer/polymer ratio and curing reaction, heterophase polymer structures are controlled by many process parameters in commercial coating compositions. The most important ones are the presence of fillers, the solvent properties and the processing conditions (application method and baking temperature). In particular, the incorporation of inorganic pigments, surface additives (polymer or oligomer surfactants), extenders or other types of fillers often interfere with the layering process. Most of these additives have a rather negative effect on the phase separation between two binders, and do not allow the development of sufficiently interesting properties compared to those of conventional multilayered systems. In many cases, liquid fillers (wetting, levelling, anti-flotting, surface active or dispersing agents) appear to impair the stratification process [8–11]. As an exception, some silicone or fluorine-containing oligomers added in minor amounts were proven to substantially accelerate the phase separation process with a moderate or negligible influence on the solutions equilibrium [9]. No previous study of the addition of a consequent amount (>2 wt.%) of a functional liquid additive in self-stratifying compositions has been reported. In this work, the aim was to design a self-stratifying coating on a polycarbonate substrate, incorporating a liquid flame retardant (FR) additive.

On the past few years, phosphorus-based flame retardants have been increasingly used as an alternative to halogen-based systems due to environmental considerations. This has generated active research in identifying novel phosphorus based FRs and also possible synergistic combinations with other elements, such as nitrogen for example [12]. Some phosphorus-containing compounds, besides their FR effect, have proven to be of great value due to their low toxicity and low smoke-emission during combustion [13]. In particular, arylphosphates have been proven to be suitable halogen-free FRs in blends containing polycarbonate (PC) (particularly in PC/ABS blends) [14,15].

In this work, film forming compositions based on a partially incompatible polymer blend, including a curable epoxy resin and a silicone resin in combination with a phosphorous-based liquid fire retardant filler are considered [16]. Two liquid fire retardant additives were tested: bisphenol-A-bis(diphenyl phosphate) (BDP) and resorcinol bis(diphenyl phosphate) (RDP). The objective was to form thin flame retardant films which spontaneously self-layer after application of one single formulation.

2. Experimental

2.1. Materials

On the basis of previous results [16], two commercial resins and a curing agent were selected, respectively: an epoxy (Bisphenol-A epoxide from Sigma-Aldrich, St. Louis, MO, USA, equivalent weight: 172–176, 100% solids), a silicone (a phenyl branched resin containing 6% of hydroxyl group from Dow Corning, Seneffe, Belgium) and a polyamine (diethylene triamine (99%), Sigma-Aldrich).

m-Xylene (99%) and butylacetate (BuAc, ≥99.5%) purchased from Sigma-Aldrich, were chosen as solvent and used without any purification.

Two phosphorus based liquid FRs, that is, bisphenol-A bis(diphenyl phosphate), Fyrolflex BDP, 8.9% of phosphorus and resorcinol bis(diphenyl phosphate), Fyrolflex RDP, 10.7% of phosphorus were purchased from ICL-Industrial Products (Amsterdam, The Netherlands) and were incorporated as additives in various amounts (2.5 wt.% and 5 wt.%).

Transparent polycarbonate plates (Lexan, 1mm-thick for microscopic analyses and 3 mm-thick for fire testing) were supplied by Polydis (Ligny Le Chatel, France).

2.2. Coatings Formulation and Application

Resins were separately dissolved at 30 wt.% in a blend of BuAc:xylene (1:1), mixed and then stirred together at a 1:1 ratio by weight (epoxy:silicone). Liquid additives were incorporated into the epoxy medium for 10 min at 300 rpm before the combination with the second resin. Finally, the curing agent was thoroughly added drop by drop, with respect to the epoxy number, and mixed for 3 min before the application of the thin film by spraying (air pressure of 200 kPa). The nominal wet film thickness was set at 200 μm. Coatings were dried at 20 °C for 24 h in an oven and then cured for 2 h at 110 °C.

2.3. Characterization of Film Properties

Microscopic analyses coupled with X-ray mappings were used to detect the potential formation of stratified layers and to determine the film thickness. Fire behavior and thermal stability of the samples were investigated using Limiting Oxygen Index (LOI), UL-94 Mass Loss Calorimetry (MLC) and ThermoGravimetric Analysis (TGA). The specimens and residues obtained were analyzed either by microscopic analyses coupled with X-ray mapping or by using a numerical microscope. To complete the evaluation of the film properties, cross-hatch testing was used to quantify the adhesion.

2.3.1. Microscopic Analyses

Coatings were cut in liquid nitrogen to allow the analysis of their cross-section and the determination of their thickness by Scanning Electron Microscopy with X-Ray mapping (SEM-EDX). Analyses were respectively carried out at 5.0 kV, 20 μA and 13.0 kV, 25 μA using a Hitachi S4700 (Tokyo, Japan) with field emission gun. Carbon metallization was carried out before any characterization.

Numerical pictures of sample residues were performed using a microscope VHX-1000 supplied by Keyence (Osaka, Japan, ×20). The microscope creates a 3D image based on automatically captured images.

2.3.2. Classification of Stratification

The degree of layering was measured following the guidelines developed in the Brite Euram Project [8]. The degree of stratification was classified from 1 to 4 according to the microscopic characterization of the cross-section of the coatings. From this ranking, a type I pattern corresponds to a perfect stratification (e.g., two well distinct and homogeneous layers), a type II pattern to an homogeneous concentration gradient between the two resins through the film thickness, a type III to the formation of spherical particles rich in one of the resins dispersed in a medium enriched with the other resin, and finally a type IV corresponds to the presence of large islands composed of a majority of one of the resins.

2.3.3. Adhesion Testing

Adhesion rating of the coating on polycarbonate was carried out following the ASTM D3359-97 standard [17]. An Elcometer 107 Cross Hatch Cutter was used to perform the test: a pressure-sensitive tape is applied and then removed over cuts made in the film, and according to the specimen obtained, adhesion is rated from 0B to 5B, 0B corresponding to the worst adhesion (no coating left on the substrate) and 5B to the best one (the edges after the test are completely smooth, and none of the squares of the lattice is detached).

2.3.4. Fire Testing

A FFT (Fire Testing Technology) mass loss calorimeter (MLC) was used to perform the experiments following the procedure defined in ASTM E906 [18]. Specimens measuring $100 \times 100 \times 3$ mm^3 were exposed in horizontal orientation to an external flux of 50 kW/m^2 and a forced ignition. The selected flux corresponds to the common heat flux generated in fully-developed fires. The equipment can be compared to the one used in oxygen consumption cone calorimetry (ASTM E-1354-90 [19]), except that a thermopile is set in the chimney in order to obtain the heat release rate (HRR) instead of using the oxygen consumption principle. The distance sample-heater was set to a distance of 35 mm from the cone base, on a ceramic backing board. The following parameters were estimated: heat release rate (HRR) as a function of time, time to ignition (TTI), peak of heat release rate (pHRR) and total heat release rate (THR). The experiments were carried out three times to ensure the repeatability of the results, and the data reported are the most representative of the three replicated experiments. Values were found to be reproducible within a relative standard deviation of $\pm10\%$.

Two vertical fire tests were used to complete the evaluation of fire performances on barrels ($100 \times 10 \times 3$ mm^3): Limiting Oxygen Index (LOI) and UL-94 test. LOI, corresponding to the minimum oxygen concentration needed to support the candle like combustion of plastics, was measured using a Fire Testing Technology instrument following the standard oxygen index test ISO 4589-2 [20]. The UL-94 test was performed according to IEC 60695-11-10 standard [21], that is, in a vertical position (the bottom of the sample is ignited with a bunsen burner). This test provides a qualitative classification of the samples, from V0 (the best ranking, when burning is short and there is no dripping of flaming particles) to NC (non-classified, i.e., with a burning of more than 30 s or up to the holding clamps at 100 mm from the ignition point).

2.3.5. Thermal Stability

Thermogravimetric Analysis (TGA)

A Discovery TGA from TA Instrument was used to carry out thermogravimetric analyses. 10 mg samples (liquid materials or dry grounded coatings samples which have undergone the same heat treatment as the coatings) were thermally decomposed in alumina crucibles under a nitrogen atmosphere. They underwent an isotherm of 120 min at 50 °C for thermal homogeneity, followed by a heating ramp from 50 to 800 °C at 20 °C·min^{-1}. The nitrogen flow rate was set at 50 mL·min^{-1}. Difference weight loss curves were calculated to detect a potential variation in the thermal stability of the systems, due to the incorporation of the filler (Equation (1)). These curves represent the difference between the experimental TG curve for the mixture ($w_{exp}(T)$) and the linear combination of TG curves ($w_{theo}(T)$) for the neat components (Equation (2)) when fillers are incorporated at x wt.%.

$$\Delta w(T) = w_{exp}(T) - w_{theo}(T) \tag{1}$$

$$w_{theo}(T) = (1-x) \times w_{resin} + x \times w_{filler} \tag{2}$$

where w_{resin} and w_{filler} correspond to the weight determined from the experimental TG curves; x the weight percentages of the sole filler. If $\Delta w(T) < 0$, then the experimental weight loss is higher than the theoretical one. This shows that the reactivity and/or interaction between the polymer and the filler leads to a thermal destabilization of the material. If $\Delta w(T) > 0$, then the system is thermally stabilized.

Heat treatments and characterization of the heat-treated residues were carried out. Heat treatments of pure components or of coatings were carried out in a tubular furnace under nitrogen flow (75 mL·min^{-1}) for 3 h at characteristic temperatures selected thanks to TG curves. The collected residues were then analyzed using a numerical microscope.

3. Results and Discussion

3.1. Stratification Study

After mixing and before application as a film, the four-component system (epoxy and silicone resins, curing agent and phosphate additive) produced homophase solutions whatever the phosphate concentration. After drying and curing, the formation of separated in-layers coating structure through the film thickness was checked and adhesion properties were evaluated (Table 1).

Table 1. Morphology and adhesion properties of the coatings containing or not phosphate additives.

Filler	wt.%	Fillers Location after Film Formation	Appearance of the Coating	Thickness (μm)	Adhesion Rating	Stratification Pattern
No filler	0	–	Slightly rough, glossy	55	5B	I
RDP	2.5	Mainly in the silicone layer, sight concentration gradient near the silicone-epoxy interface	Less rough than without additive, higher gloss	45	5B	II-III
RDP	5	Mainly in the silicone layer, sight concentration gradient near the silicone-epoxy interface	Less rough than without additive, higher gloss	60	5B	I
BDP	2.5	High concentration in the silicone layer, concentration gradient in the epoxy phase from top to bottom	Less rough than without additive, higher gloss	68	4B	II-III
BDP	5	High concentration in the silicone layer, concentration gradient in the epoxy phase from top to bottom	Less rough than without additive, higher gloss	30	4B	I

The reference system without any phosphate additive shows perfect stratification (type I pattern) and the best adhesion rating (5B) on polycarbonate: the hydrophobic layer (the silicone medium) migrated to the upper part of the coating whereas the epoxy system was found at the interface with the substrate.

In presence of liquid FR fillers, nice coatings free of bubbles or defaults were obtained in all cases. They even show slight improvement in terms of surface aspect (less rough and glossier) than the reference coating. The 5B adhesion rating is maintained after the incorporation of RDP in the system, for both amounts tested, whereas the addition of BDP slightly impacts the adhesion of the coating (4B rating is obtained).

Unlike the results mentioned in the literature, it appears that neither RDP nor BDP affect the layering process when incorporated at 5 wt.% in the formulation: a type I pattern is obtained with fully separated layers, as observed with the unfilled system (Figure 2).

Both additives migrate to the upper layer of the film, in the silicone medium, meaning that RDP and BDP have a higher affinity with silicone than with epoxy. This migration is rather a positive phenomenon in this case as it allows concentrating the additives' properties (i.e., flame retardant properties) in the upper part of the coating, where it is the most needed in case of fire. With RDP, some inhomogeneities are noticeable along the film near the two resins interface region: some island shaped regions rich in silicone have not totally migrated to the upper layer of the film. A concentration gradient of phosphorus is also detectable through the film thickness, although the major part of RDP has migrated to the upper layer. With BDP, the same behavior can be observed: the silicone resin and the filler are mainly concentrated in the upper part of the coating. However, the concentration gradient of phosphorus is more pronounced through the thickness than with RDP. The difference could be explained by a higher affinity of RDP with the silicone resin, promoting the migration of the additive to the air interface. However, the concentration gradient is much more noticeable with phosphorus than with silicium: this means that the affinity between the silicone resin and the filler is not strong enough to allow its complete migration toward the top layer. A difference in viscosity could also explain the difference in concentration gradient between BDP and RDP containing formulations. Indeed, BDP is much more viscous than RDP (17,000 CPS [22] compared to 12,450 CPS for RDP [23]. As its viscosity is higher, its migration to the upper layer may be slowed down. It is noteworthy to notice that even if the additive has not enough time to completely migrate to the upper layer of the film, it does not impair the stratification process: the layering of the silicone to the top of the coating remains perfect.

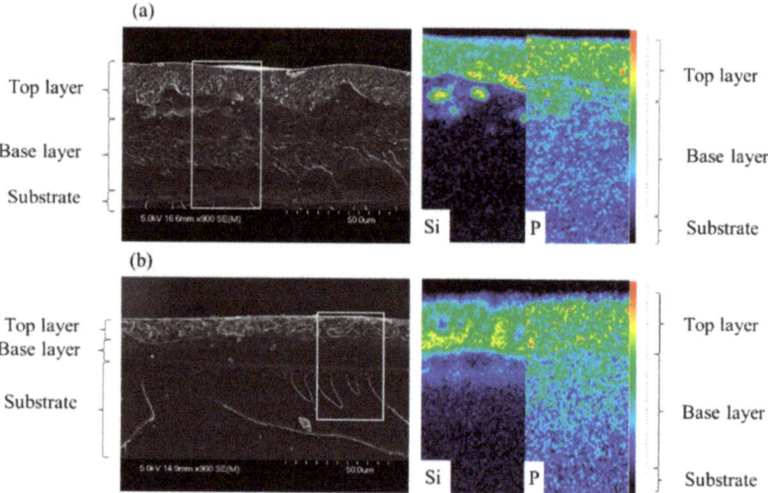

Figure 2. EDS X-ray mappings of Silicon and Phosphorus on a cross-section of an epoxy/silicone coating filled with 5 wt.% of (**a**) resorcinol bis(diphenyl phosphate) (RDP) and (**b**) bisphenol-A-bis(diphenyl phosphate) (BDP) additives.

When a smaller amount of phosphate additives is introduced in the formulation (2.5 wt.%), evidences of phase separation are found with a concentration gradient of silicone through the thickness. Nevertheless, some inhomogeneities in the matrix are observed (Figure 3), consisting in isolated spherical particles composed of epoxy resin dispersed in the continuous silicon matrix. Furthermore, the phosphorus element is found in the silicone phase whereas the stratification is not complete in those areas. This corroborates its higher affinity with silicone compare to that of epoxy. The phosphate additive destabilizes the interface between the two phases and, in that case, influences the preferential orientation of the phases in the course of film forming process. The dispersion of the filler is portioned in the silicone medium, however interacting in some way in the course of crosslinking reactions. A minimum amount of filler may be needed to allow a homogeneous dispersion into the silicone medium and this would need further investigation to the forces questioned in this process. This morphology corresponds to a type II/type III stratification pattern. However, this incomplete phase separation has no influence on the adhesion properties of the coatings as similar adhesion ratings are obtained with 2.5 wt.% and 5 wt.% of fillers, for both RDP and BDP [24].

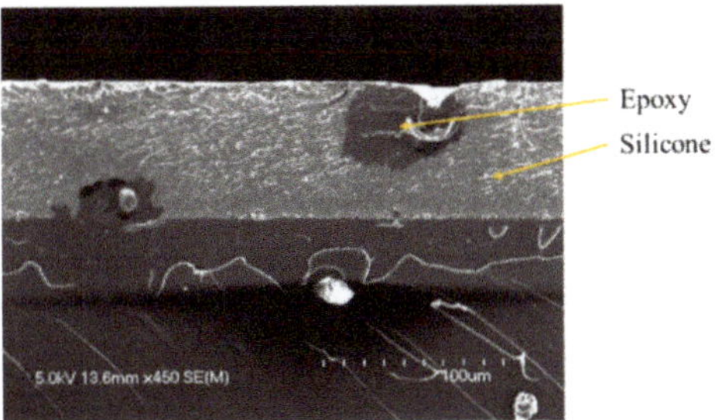

Figure 3. Scanning Electron Microscopy (SEM) pictures of the cross-section of the ternary system filled with 2.5 wt.% of BDP.

In conclusion, when a low amount of additive is added (2.5 wt.%), self-layering is slightly compromised in some areas of the film. The filler destabilizes the equilibrium between the two solutions, and impairs the formation of two distinct and homogeneous layers. On the contrary, the phosphorus-based fillers tested do not affect the formation of oriented heterophase structures when they are introduced at 5 wt.%: they mostly migrate, with a concentration gradient more or less pronounced, towards the air interface with the silicone resin. In that case, some disparities in the concentration gradient can be observed, which may be due to a difference in viscosity between the two phases. Finally, coating's composition provided nice visual appearance and a high adhesion onto the polycarbonate substrate. It was proven that it is possible to add a substantial amount (5 wt.%) of liquid functional filler in a self-layering formulation without affecting the self-stratification process, the coating aspect and the adhesion. In these formulations, the phosphorus FR additives migrate into the upper silicone layer, which can potentially lead to an effective FR effect of the coating when exposed to fire.

3.2. Flame Retardant Properties

Virgin PC and PC coated with the unfilled and filled self-stratified compositions were then tested for fire performances (MLC) and flammability (UL-94 and LOI).

HRR curves and characteristic MLC parameters of the systems are given in Figure 4 and Table 2. It is noteworthy that no ignition of the coated samples (filled and unfilled) occurs at 35 kW/m^2 (mild fire scenario), even after 30 min of exposure; whereas the raw PC ignites after 319 s and releases a total heat of 35 MJ/m^2 (with a pHRR of 202 kW/m^2, see Supplementary Materials Figure S1). PC is a char-forming polymer. At 50 kW/m^2, after its ignition, it melts and forms a char which swells until a height of about eight centimeters is reached. The char then degrades and only ashes remain at the end of the test.

Even at a higher heat flux (50 kW/m^2), the coating allows protecting temporarily the PC. MLC results obtained for the coating with the unfilled epoxy/silicone coating evidence an improvement of the fire behavior, as both the pHRR and THR are reduced, respectively by 24% and 21% compared to raw PC. Ignition time is also delayed more than 50 s, which is remarkable with such thickness (55 μm). Ignition occurs only because PC swells under the coating, which thus progressively delaminates from the edges of the plate due to the uprising of the PC char. Once ignited, PC forms a char which rapidly swells up to 8 cm. The char is then consumed and collapses before the flame out takes place. At the end of the test, PC is completely consumed, which also explains the pHRR and THR values obtained.

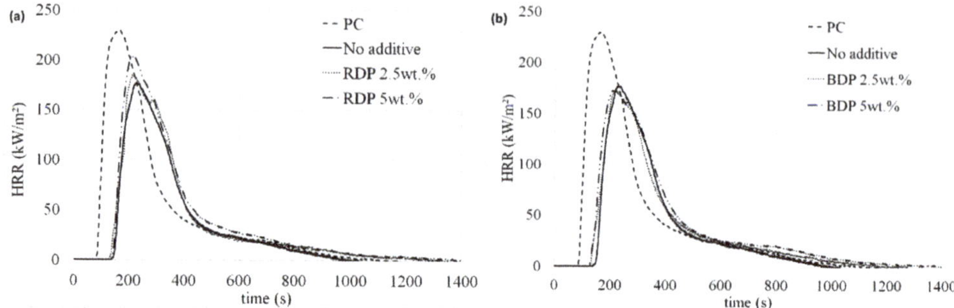

Figure 4. Heat release rate (HRR) curves obtained with uncoated polycarbonate (PC) and coated PC with the self-stratified epoxy/silicone compositions containing (**a**) RDP and (**b**) BDP flame retardant (FR) additives at 2.5 wt.% and 5 wt.%.

Table 2. Mass loss calorimeter (MLC) values obtained at 35 and 50 kW/m with uncoated and coated PC with the epoxy/silicone composition with and without FR additives.

Heat Flux, Parameters Measured		Raw PC	No Additive	With RDP		With BDP	
				2.5 wt.%	5 wt.%	2.5 wt.%	5 wt.%
35 kW/m^2	TTI (s)	319	No ignition	No ignition		No ignition	
	pHRR (kW/m^2)	202					
	THR (MJ/m^2)	35					
50 kW/m^2	TTI (s)	92	148	154	143	140	132
	pHRR (kW/m^2)	231	176	189 (+7%) [1]	209 (+19%)	174 (−1%)	180 (+2%)
	THR (MJ/m^2)	52	41	45 (+9%)	50 (+21%)	44 (+7%)	50 (+21%)

[1] Percentages represent the difference compared to the unfilled epoxy/silicone system.

MLC experiments on PC samples coated with only one resin, whether the epoxy resin or the silicone one, have been carried out to try to explain the behavior of PC coated with the unfilled epoxy/silicone blend (Figure 5, Table 3). When the epoxy resin is applied on PC, the behavior during the MLC test is similar to that of pure PC: no delay of the TTI and similar pHRR and THR. Thus, the epoxy resin does not bring any fire retardant effect, which is not surprising as it is composed of Bisphenol A (similarly to PC). On the contrary, the application of a silicone coating on PC leads to a fire retardant effect: slight decrease of pHRR and THR and TTI increase of about 40 s. Thus, the presence of silicone is responsible for the improvement of the fire behavior when the epoxy/silicone coating is applied to PC. However, the epoxy resin contributes to bringing better adherence between the self-stratifying coating and PC. In fact, the silicone coating gives a 3B rating on PC compared to a 5B rating with the self-stratifying coating. Moreover, there is no interlayer adhesion failure between the epoxy and the silicone layers thanks to the use of the self-stratifying coating.

Similar trends are observed when PC is coated with the epoxy/ silicone/ filler mixture, whatever the amount of additive incorporated. However, although the best TTI is obtained when RDP/BDP are incorporated at 2.5 wt.%, fire behavior are similar compared to that of the binary composition. The impact of RDP and BDP addition in terms of flame retardant behavior is thus negligible compared to the FR effect of the silicone coating itself. Finally, BDP seems slightly more efficient than RDP, which is not surprising as at a same ratio, BDP contains more phosphorus than RDP. Indeed, there is 10.7% of P in RDP whereas 8.9% are present in BDP. All the systems were consumed at the end of the test at 50 kW/m^2: only residue of silica and ashes remained. The slight thickness difference between the two samples (Table 1) is not responsible for this effect since any correlation is noted between the thickness, the TTI and THR: close THRs are reached by both filled coatings at equal weight percentage whereas a reverse effect is observed on the time before ignition.

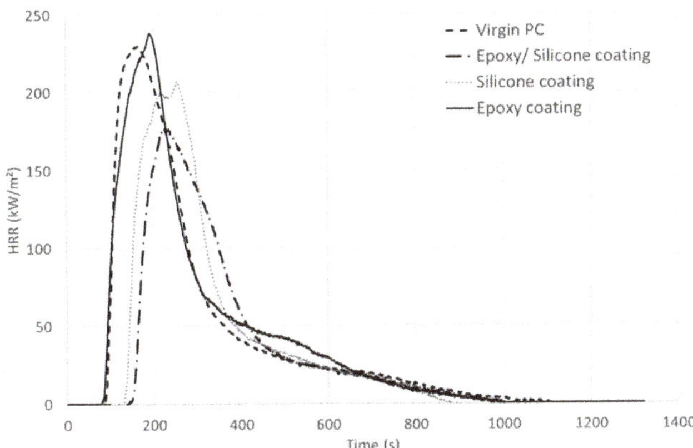

Figure 5. HRR curves obtained with uncoated PC and coated PC with the self-stratified epoxy/silicone coating, the epoxy coating and the silicone coating alone.

Table 3. MLC values obtained at 50 kW/m^2 with uncoated and coated PC with the epoxy/silicone self-stratifying coating, the epoxy and the silicone coating solely.

Parameters Measured	Uncoated PC	Coated PC		
		Epoxy/Silicone Self-Stratifying Coating	Epoxy Coating	Silicone Coating
TTI (s)	92	148	80	131
pHRR (kW/m^2)	231	176 (−24%) [1]	238 (+3%)	206 (−11%)
THR (MJ/m^2)	52	41 (−21%)	53 (+2%)	45 (−13%)

[1] Percentages represent the difference compared to uncoated PC.

LOI and UL-94 tests were also performed on uncoated and coated PC formulations (Table 4). The epoxy/silicone reference coating does not improve the intrinsic flame retardant properties of PC in the case of vertical burning tests: close LOI values (27 vol.% and 28 vol.% respectively for the raw and coated PC) and a NC rating are obtained for both samples. Up to 28 vol.%, the extinction of virgin PC occurs mainly because of the flaming drops or of the rapid combustion of the material. The unfilled coating allows reducing the combustion speed of the material and prevents the dripping of PC. At UL-94 test, virgin PC is not classified at 3 mm, but close to meet the requirements of V-2 classification: short combustion time (<32 s) and dripping (Table 5).

When the polymer is coated with the filled mixture, the fire behavior is strongly dependent on the amount of fillers used in the coating, whatever the type of additive (BDP or RDP). The addition of 2.5 wt.% of additives does not improve the fire behavior compared to the unfilled system: similar LOI value and same rating at UL-94 test. For these materials, the amount of additives is too low to allow the decrease of the flame spread. With 5 wt.% of additives, although the rating at UL-94 test is the same (NC), the behavior is much different: the combustion time is very short (≤2 s) after the first ignition and remains shorter than the unfilled system after the second ignition (Table 5). The classification is very close to V-1, as with both additives= only one specimen out of the five tested has a t$_2$ afterflame time higher than 30 s (in bold, Table 5). In addition, LOI values are significantly improved: 34 vol.% and 32 vol.% respectively with RDP and BDP fillers. The charring is more pronounced (fast expansion of the foamed structure) and no dripping occurs. Last but not least, samples were more prone for burning during the second ignition with a lower concentration of additive in the coating.

Thus, similar behavior between RDP and BDP can be registered: they are both more efficient when they are introduced at a higher amount (5 wt.%). As with MLC, the slight difference observed between RDP and BDP may be due to the higher amount of phosphorus in the RDP formulation.

Table 4. LOI, UL-94 rating for PC, and coated PC with the binary and ternary mixtures.

Fire Test	Raw PC	No Additive	With RDP		With BDP	
			2.5 wt.%	5 wt.%	2.5 wt.%	5 wt.%
LOI (vol.% O_2)	27	28	27	34	26	32
UL-94	NC Dripping	NC No dripping	NC No dripping	NC No dripping	NC No dripping	NC No dripping

Table 5. Time of the two flaming combustions during UL-94 test.

Sample	Raw PC	No Additive	With RDP		With BDP	
			2.5 wt.%	5 wt.%	2.5 wt.%	5 wt.%
Ignition	1st	2nd	1st	2nd	1st	2nd
Flaming combustion time	24	15	16	90	28	36
	32	17	6	7	19	5
	18	23	6	47	27	80
	11	6	2	50	4	13
	23	5	11	5	42	20
Total flaming time	108	66	95	199	120	154

To conclude, formulations containing 5 wt.% of RDP and BDP exhibit the most promising results: charring is more pronounced and dripping is avoided. At MLC test, fire performances of coated formulations are mainly enhanced through the shift of the TTI toward higher values due to the silicone coating. The influence of the two phosphorus-based additives on the fire properties is not significant compared to the improvements already reached by the unfilled epoxy/silicone system.

Occurrence and efficiency of the systems depend not only on the FR itself, but also on the interactions with the degradation products of the different components of the materials (resins and substrate). To go deeper in the understanding of the mode of action of the phosphorus compounds, TG analyses and difference weight loss calculations were carried out on the silicone resin, containing phosphorus additives or without. During a fire scenario, the silicone will be the resin in contact with the open flame or exposed to the heat source as it is located in the upper layer of the film. Consequently, only the silicone resin was considered.

3.3. Thermal Stability

The thermal stability of the unfilled system and systems filled with RDP and BDP were investigated using TG analyses (Figures 6 and 7. Weight difference curves between experimental and calculated TG curves when RDP or BDP is added to the resin (10 wt.%) under N_2 conditions). The percentage of fillers introduced in the silicone medium was kept constant compared to the epoxy/silicone system as the major part of the filler migrates to the silicone layer (i.e., 10 wt.% of filler in the silicone medium).

The decomposition of the pure silicone film involves a three-step process with a 68 wt.% residual weight at 800 °C. The first degradation step may be coupled with the release of the remaining solvents from the resin's preparation and the release of silicone oligomers (11 wt.%) and the second and third steps (overlapped) are correlated with to the release of aromatic compounds (such as benzene and bisphenyl, 21 wt.%) [25]. The total weight loss is only 32% at 800 °C, demonstrating the excellent thermal stability of the silicone resin (under TGA conditions). When FR liquid additives are incorporated, an additional degradation step occurs close to the maximum degradation temperature of RDP and BDP (at respectively 398 and 420 °C). This leads to a slight destabilization of the system, particularly with BDP between 370 and 480 °C (−2 %/°C, Figure 7). Weight difference curves between experimental and calculated TG curves when RDP or BDP is added to the resin (10 wt.%) under N_2 conditions). The residual weights of the additives are low (respectively 4 wt.% and 17 wt.% at 800 °C with BDP and RDP), although it does not neither influence significantly the thermal stability of the resins nor favor the formation of a charring structure at high temperature.

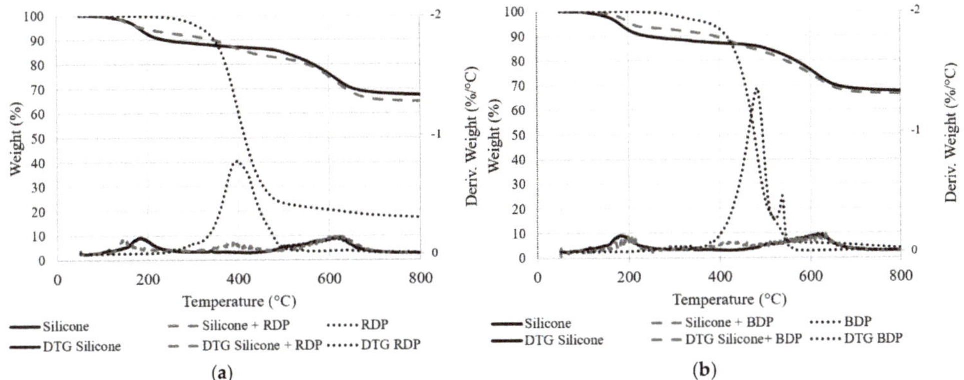

Figure 6. Comparison of TG and DTG curves of the silicone system with and without (**a**) RDP and (**b**) BDP (10 wt.%) under nitrogen at a heating rate of 20 °C/min.

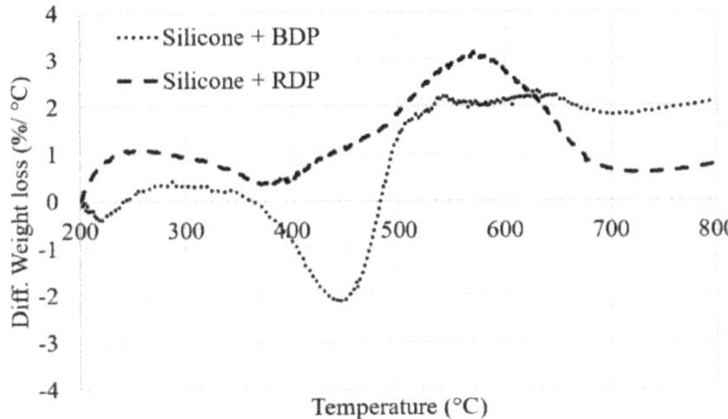

Figure 7. Weight difference curves between experimental and calculated TG curves when RDP or BDP is added to the resin (10 wt.%) under N_2 conditions.

TG experiments enable us to define characteristic temperatures of degradation. Accordingly, heat treatments were performed at 300 °C in a tubular furnace and under a nitrogen flow on the silicone and the silicone/fillers systems (Figure 8). From the numerical pictures, the formation of an expanded foamed structure with small cells, whatever the phosphorus compound used is evidenced: the fillers do not modify the cellular structure. Finally, RDP and BDP have low influence on the thermal stability of the silicone resin; the amount of the remaining residue at 800 °C is almost the same, and the degradation process remains unchanged compared to the pure silicone. Based on those observations, the formation of an expanded coating during the test explains, at least partially, the fire protection brought by the paint and the silicone-based coating containing the fillers. The silicone and silicone/filler compositions allow the formation of a protective barrier, thus reducing heat transfers from the external heating source to the substrate.

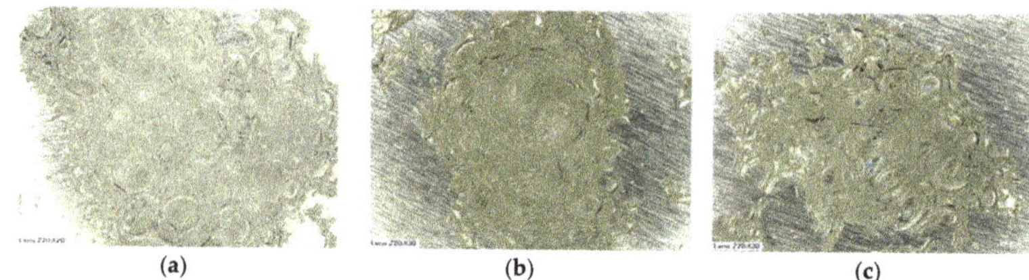

(a)   (b)   (c)

Figure 8. Numerical pictures of residues of (**a**) silicone, and (**b**) silicone/RDP, (**c**) silicone/BDP at 10 wt.% at 300 °C.

Visual observations during the horizontal fire tests confirm the higher char yield obtained when phosphorus compounds are incorporated into the systems, even if thermal analyses did not show major improvements in terms of stability. Based on the literature review, phosphorus compounds accumulate in the condensed phase in the char layer, at the surface of the burning specimens in a blend of PC–ABS + RDP [26]. Results suggested that PC undergoes photo-fries rearrangement upon thermal decomposition, and RDP would react with the formed phenolic groups through a trans-esterification mechanism. Kinetic analysis of the thermal decomposition of PC containing RDP supported the proposed mechanism.

In conclusion, the high thermal stability of the silicone resin under inert atmosphere at elevated temperatures has been demonstrated. As this component is at the interface with the air in the self-stratifying coating composition, it will consequently be the first in contact with either the heat source or the flame during the fire tests. According to the results, the silicone resin provides some thermal stability to the system, which may explain the higher time to ignition obtained during MLC experiments. Under elevated temperatures, the silicone coating protects the underlying substrate (and the epoxy resin) from decomposition and creates a barrier between the heat source and the substrate. Moreover, the similar behavior in terms of thermal stability for the silicone resin with and without fillers could explain the close results at MLC test for all the epoxy/silicone coatings.

4. Conclusions

A properly-selected partially incompatible polymer blend composed of silicone, of a curable epoxy resin and of a liquid functional filler successfully formed a double-layered coating, showing excellent adhesion to the underlying polycarbonate substrate. The binders, dissolved in solvents to produce a two-phase liquid medium during film formation, led to structures with sharply defined layers. The topcoat layer was found to be composed of the silicone resin, and the base layer of the curable epoxy resin. Microscopic analyses demonstrated that phosphorus-based liquid fillers do not impact the stratification process when incorporated at 5 wt.%. However, their migration to the upper layer of the coating (silicone phase) was not always complete: a concentration gradient of additives through the film thickness is obtained, with a higher concentration toward the top of the coating. When fillers were incorporated at a lower amount (2.5 wt.%), incomplete phase separation was observed in some areas of the film, although a high adhesion rating (5B and 4B with RDP and BDP fillers respectively) and nice visual appearance remained.

The two phosphorus-based compounds led to an increase in the LOI value: up to 34 vol.% is obtained with 5 wt.% RDP compared to 27 vol.% for the unfilled system. However, the systems are still non classified at UL-94, even if combustion times with 5 wt.% additives are very close to a V1 rating. For MLC test at 35 kW/m^2, the coated samples do not ignite, and at 50 kW/m^2, the application of the epoxy/silicone coating allows delaying the TTI, with and without the presence of phosphate additives: improvements are due to the silicone resin and are not significantly modified by the addition of the liquid fire retardants. Synergistic effects between the two liquid fillers are not expected, the level of flame properties reached are mainly dependent on the amount of phosphorus introduced in the system.

Finally, numerous factors are of importance for the formation of two well separated layers. It is known that the efficiency and mechanism of action of phosphorus flame retardants in coating compositions are influenced and can be optimized by modifying them using specific synergists. In that respect, the surface tension of the paint components, their viscosity and the amount introduced in the coating compositions have a decisive role.

Supplementary Materials: The following are available online at http://www.mdpi.com/2079-6412/8/12/448/s1, Figure S1: HRR curves obtained with PC and coated PC with the self-stratified epoxy/silicone compositions containing RDP and BDP FR additives at 2.5 wt. % and 5 wt. % (the four curves overlapped).

Author Contributions: Conceptualization, A.B., M.J., M.C., S.B., S.D. (Stephanie Degoutin), S.D. (Sophie Duquesne), C.P. Methodology, A.B., M.J., S.B., C.P.; Software, S.B., C.P.; Validation, M.J., M.C., S.B., S.D. (Stephanie Degoutin), S.D. (Sophie Duquesne), C.P.; Investigation, A.B., S.B.; Resources, M.J., M.C., S.B., S.D. (Stephanie Degoutin), S.D. (Sophie Duquesne), C.P.; Data Curation, A.B.; Writing-Original Draft Preparation, A.B.; Writing-Review & Editing, M.J., M.C., S.D. (Stephanie Degoutin), S.D. (Sophie Duquesne), C.P.; Visualization, A.B., M.J., M.C.; Supervision, M.J., M.C. and S.D. (Sophie Duquesne); Project Administration, M.J.; Funding Acquisition, M.J.

Funding: This research was funded by the french ANR (Agence Nationale de la Recherche, No. ANR-14-CE27-0010), STIC project.

Acknowledgments: We are also thankful to the MATIKEM competitiveness cluster for supporting the project and Lille University for administrative support.

Conflicts of Interest: The authors declare no conflict of interest.

References

1. Novak, B.M. Hybrid nanocomposite materials—Between inorganic glasses and organic polymers. *Adv. Mater.* **1993**, *5*, 422–433. [CrossRef]

2. Mackulin, P.J.; Krafcik, R.B. Hybrid Latex Particles For Self-Stratifying Coatings. U.S. Patent 20140303281 A1, 9 October 2014.

3. Wen, J.; Jordens, K.; Wilkes, G.L. Hybrid organic/inorganic coatings for abrasion resistance on plastic and metal substrates. *MRS Online Proc. Lib. Arch.* **1996**, *435*, 207. [CrossRef]

4. Yagci, M.B.; Bolca, S.; Heuts, J.P.A.; Ming, W.; de With, G. Self-stratifying antimicrobial polyurethane coatings. *Prog. Org. Coat.* **2011**, *72*, 305–314. [CrossRef]

5. Prolongo, S.G.; Moriche, R.; Sánchez, M.; Ureña, A. Self-stratifying and orientation of exfoliated few-layer graphene nanoplatelets in epoxy composites. *Compos. Sci. Technol.* **2013**, *85*, 136–141. [CrossRef]

6. Funke, W. Preparation and properties of paint films with special morphological structure. *J. Oil Colour Chem. Assoc.* **1976**, *59*, 398–403.

7. Murase, H.; Funke, W. Formation of two-phase coatings from a mixture of powdered polymers. In *Congress Book, XVth FATIPEC Congress, 8–13 June 1980, Amsterdam: 3E-Activities*; Netherlands Association of Coatings Technologists NVVT: Amsterdam, The Netherlands, 1980; Volume 2, pp. 387–409.

8. Toussaint, A. Self-stratifying coatings for plastic substrates (Brite euram project RI 1B 0246 C(H)). *Prog. Org. Coat.* **1996**, *28*, 183–195. [CrossRef]

9. Verkholantsev, V.; Flavian, M. Polymer structure and properties of heterophase and self-stratifying coatings. *Prog. Org. Coat.* **1996**, *29*, 239–246. [CrossRef]

10. Verkholantsev, V.; Flavian, M. Epoxy thermoplastic heterophase and self-stratifying coatings. *Mod. Paint Coat.* **1995**, *85*, 100–107.

11. Vink, P.; Bots, T.L. Formulation parameters influencing self-stratification of coatings. *Prog. Org. Coat.* **1996**, *28*, 173–181. [CrossRef]

12. Joseph, P.; Ebdon, J.R. Phosphorus-based flame retardants. In *Fire Retardancy of Polymeric Materials*, 2nd ed.; Wilkie, C.A., Morgan, A.B., Eds.; CRC Press: Boca Raton, FL, USA, 2010; pp. 107–127.

13. Yeh, J.T.; Hsieh, S.H.; Cheng, Y.C.; Yang, M.J.; Chen, K.N. Combustion and smoke emission properties of poly(ethylene terephthalate) filled with phosphorous and metallic oxides. *Polym. Degrad. Stab.* **1998**, *61*, 399–407. [CrossRef]

14. Dressler, H. The uses of resorcinol/derivatives in polymers. In *Resorcinol: Its Uses and Derivatives*; Springer: Boston, MA, USA, 1994.

15. Price, D.; Anthony, G.; Carty, P. Introduction: Polymer combustion, condensed phase pyrolysis and smoke formation. In *Fire Retardant Materials*; Horrocks, A.R., Price, D., Eds.; CRC press: Cambridge, UK, 2001; pp. 1–30.

16. Beaugendre, A.; Degoutin, S.; Bellayer, S.; Pierlot, C.; Duquesne, S.; Casetta, M.; Jimenez, M. Self-stratifying epoxy/silicone coatings. *Prog. Org. Coat.* **2017**, *103*, 101–110. [CrossRef]

17. *ASTM D3359-97 Standard Test Methods for Measuring Adhesion by Tape Test*; ASTM International: West Conshohocken, PA, USA, 1997.

18. *ASTM E-906 Standard Test Method for Heat and Visible Smoke Release for Materials and Products*; ASTM International: West Conshohocken, PA, USA, 1987; pp. 921–942.

19. *ASTM E-1354-90 Standard Test Method for Heat and Visible Smoke Release for Materials and Products Using an Oxygen Consumption Calorimeter*; ASTM International: West Conshohocken, PA, USA, 1990.

20. *ISO 4589-2 Determination of Burning Behavior by Oxygen Index*; International Organization for Standardization: Geneva, Switzerland, 2017.

21. *IEC 60695-11-10 Fire Hazard Testing—Part 11-10: Test Flames—50W Horizontal and Vertical Flame Test Methods*; International Electrotechnical Commission: Geneva, Switzerland, 2013.

22. ICL Industrial Product, Fyrolflex® BDP, Bis-phenol A-bis(diphenyl phosphate). Available online: http://icl-ip.com/wp-content/uploads/2012/05/090625_Fyrolflex_BDP.pdf (accessed on 5 December 2018).

23. ICL Industrial Product, Fyrolflex™ RDP, Resorcinol bis(diphenyl phosphate). Available online: http://icl-ip.com/products/fytol-rdp/ (accessed on 5 December 2018).

24. Verkholantsev, V.V. Nonhomogeneous-in-layer coatings. *Prog. Org. Coat.* **1985**, *13*, 71–96. [CrossRef]

25. Gardelle, B.; Duquesne, S.; Vu, C.; Bourbigot, S. Thermal degradation and fire performance of polysilazane-based coatings. *Thermochim. Acta* **2011**, *519*, 28–37. [CrossRef]

26. Murashko, E.A.; Levchik, G.F.; Levchik, S.V.; Bright, D.A.; Dashevsky, S. Fire-retardant action of resorcinol bis(diphenyl phosphate) in PC–ABS blend. II. Reactions in the condensed phase. *J. Appl. Polym. Sci.* **1999**, *71*, 1863–1872. [CrossRef]

 *coatings*

Article

Bio-Inspired Fluorine-Free Self-Cleaning Polymer Coatings

Lionel Wasser [1], Sara Dalle Vacche [1,2], Feyza Karasu [1,†], Luca Müller [1], Micaela Castellino [2,3], Alessandra Vitale [2], Roberta Bongiovanni [2] and Yves Leterrier [1,*

[1]   Laboratory for Processing of Advanced Composites (LPAC), Ecole Polytechnique Fédérale de Lausanne (EPFL), CH-1015 Lausanne, Switzerland; lionel.wasser@gmail.com (L.W.); sara.dallevacche@polito.it (S.D.V.); feyza.karasukilic@unifr.ch (F.K.); luca.mueller@epfl.ch (L.M.)

[2]   Department of Applied Science and Technology Politecnico di Torino, Corso Duca degli Abruzzi 24, 10129 Turin, Italy; micaela.castellino@polito.it (M.C.); alessandra.vitale@polito.it (A.V.); roberta.bongiovanni@polito.it (R.B.)

[3]   Center for Sustainable Future Technologies, CSFT, IIT@Polito, Italian Institute of Technology, Corso Trento 21, 10129 Turin, Italy

*   Correspondence: yves.leterrier@epfl.ch; Tel.: +41-21-693-4848

†   Present address: Adolphe Merkle Institute, University of Fribourg, Chemin des Verdiers 4, 1700 Fribourg, Switzerland.

Received: 30 September 2018; Accepted: 26 November 2018; Published: 28 November 2018

Abstract: Bio-inspired fluorine-free and self-cleaning polymer coatings were developed using a combination of self-assembly and UV-printing processes. Nasturtium and lotus leaves were selected as natural template surfaces. A UV-curable acrylate oligomer and three acrylated siloxane comonomers with different molecular weights were used. The spontaneous migration of the comonomers towards the polymer–air interface was found to be faster for comonomers with higher molecular weight, and enabled to create hydrophobic surfaces with a water contact angle (WCA) of 105°. The replication fidelity was limited for the nasturtium surface, due to a lack of replication of the sub-micron features. It was accurate for the lotus leaf surface whose hierarchical texture, comprising micropapillae and sub-micron crystalloids, was well reproduced in the acrylate/comonomer material. The WCA of synthetic replica of lotus increased from 144° to 152° with increasing creep time under pressure to 5 min prior to polymerization. In spite of a water sliding angle above 10°, the synthetic lotus surface was self-cleaning with water droplets when contaminated with hydrophobic pepper particles, provided that the droplets had some kinetic energy.

Keywords: self-cleaning; lotus; nasturtium; siloxane surfactants; acrylates; photopolymerization; UV nanoimprint lithography; PDMS template

1. Introduction

Natural superhydrophobic self-cleaning surfaces such as the famous lotus leaf consist of an intrinsic hierarchical structure with epithelial cells of sizes in the micrometer range, and sub-micron, low surface energy epicuticular wax crystals [1,2]. Such structures favor trapping of small air pockets at the interface with water droplets, which considerably reduces the contact area between the droplet and the surface, resulting in the reduction of contact angle hysteresis, tilt angle, and adhesive force. The self-cleaning effect is the removal of dirt particles by the (rain) water, which form droplets that do not 'stick' on the surface and run away with the dirt. The development of bio-inspired synthetic self-cleaning surfaces has stimulated a considerable research effort since more than a decade, leading to remarkable results. However, reported surfaces often rely on environmentally-detrimental and cost-intensive approaches. These are primarily based on fluorinated compounds [3] owing to their low surface energy, and to nanoparticles giving rise to nanopatterned structures: some fluorinated compounds—i.e., long chained fluoroalkylic compounds—create concerns for their biopersistency, while nanoparticles are possibly associated with uncontrolled release issues [4]. In addition, many of the demonstrated methods are hardly scalable to cost-effective, large area surfaces [5,6]. Efforts to create fluorine-free and scalable self-cleaning surfaces have been initiated in the last few years, including superhydrophobic water-based nanoparticulate dispersions [7,8], sprayed waxes dissolved in supercritical CO_2 [9], and various micro/nanotextures impregnated with lubricating liquids [10,11]. The present work follows up with the development of low-energy surfaces based on the spontaneous, enthalpy-driven migration of comonomers and resulting segregation at the polymer–air interface [12–16]. It is based on a highly scalable low-pressure, solvent-free, and ambient UV replication process of plant surfaces using a silicone template as demonstrated in a recent work [17]. In this study, one of the main challenges was the reverse migration of the fluorinated surfactant comonomer, from the polymer–air interface back to the bulk, upon contacting the low-surface energy silicone template, and suppression of the superhydrophobic properties. This problem was solved using a flash of UV light to chemically attach the comonomer previously segregated at the polymer surface, prior to UV printing, but this increased the viscosity of the superficial polymer layers, which was detrimental to the low-pressure replication fidelity.

The objective of this work was to produce bio-inspired self-cleaning coatings using siloxane surfactant comonomers as fluorine-free alternatives, and a cost-effective, low-pressure UV printing process. The leaves of two superhydrophobic plants—namely lotus and nasturtium—were selected to create templates. These two plants possess similar epicuticular wax crystals in the form of sub-micron tubules, however they exhibit totally different hierarchical microstructures: the lotus leaf structure is composed of microscale papillae, whereas nasturtium has larger convex epidermal cells [18,19]. Attention was also paid to the influence of the molecular weight and concentration of Si in the surfactant comonomer, and on the reverse migration phenomenon on the superhydrophobic character of synthetic replica of plant surfaces.

2. Materials and Methods

2.1. Materials

A hyperbranched polyester acrylate oligomer (CN2302, Sartomer, Colombes, France) was selected owing to its low polymerization shrinkage [20], which warrants a very high replication fidelity [21–23]. It has a theoretical functionality of 16 and a real one of 13, a density of 1.13 $g \cdot cm^{-3}$ and a Newtonian viscosity of 0.3 Pa s at 25 °C. The photo-initiator was diphenyl(2,4,6-trimethylbenzoyl) phosphine oxide (TPO Esacure, Lamberti, Gallarate, Italy). Three different siloxane comonomers were used, namely an acryloxy-terminated polydimethylsiloxane (PDMSat, ABCR GmbH, Karlsruhe, Germany), and two polyether-modified polysiloxane polyurethane acrylates (PESiUA1 and PESiUA2) whose synthesis and structures are described elsewhere [24]. Table 1 provides molecular weight, number of repeating siloxane units, and amount of Si in the three comonomers.

Table 1. Molecular weight, atomic and mass concentrations of Si for the siloxane comonomers.

Siloxane Comonomer	Molecular Weight (g/mol)	Number of –(Si–O)– Units	Concentration of Si (wt %)
PDMSat	581	3	14.5
PESiUA1	5600	16	8.7
PESiUA2	10400	16	4.3

A total of 12 formulations were prepared with the three comonomers, at concentrations equal to 0.5, 1, 2, and 5 wt %. For each formulation, the hyperbranched acrylate was first mixed with a concentration of 6 wt % of TPO at 75 °C and stirred for 15 min using a magnetic stirrer, until the mixture was homogenous. The hyperbranched acrylate + TPO was used as the control material and will be referred to as the acrylate in the following. PDMSat formulations were prepared by mixing selected PDMSat amounts with the acrylate at 40 °C for further 15 min. PESiUA1 and PESiUA2 formulations were more difficult to prepare due to the very high viscosity of these two comonomers, which prevented accurate dosage and homogeneous mixing with the acrylate. To overcome this problem, the comonomers were heated for 10 min at 50 °C and selected amounts could be mixed with the acrylate and stirred for at least 12 h at ambient temperature. All formulations were transparent and stirred again prior to further processing in order to guarantee a good homogeneity.

2.2. Process Methods

Flat surfaces were produced in a first step in order to study the migration of siloxane oligomers. 200 μm thick coatings were prepared on glass slides using a doctor blade and photopolymerized immediately, or after selected times to allow monomer migration to the free surface exposed to air. Curing was then performed using a 200 W high-pressure mercury lamp (OmniCure 2000, EXFO, Mississauga, ON, Canada) and a collimator positioned at 12 cm above the sample during 3 min under a UV intensity of 75 mW·cm^{-2} at the surface of the sample. The light intensity at the surface of the sample was measured between 230 and 410 nm using a calibrated radiometer (Silver Line, CON-TROL-CURE, Chicago, IL, USA). Notice that for the surfaces photopolymerized under air, the concentration of photoinitiator was high enough to overcome oxygen inhibition, and the cured surfaces were non-sticky and hard. This was further checked by measuring the water contact angle (WCA): Surfaces polymerized under N_2 and under air showed the same wettability (see Table 2).

Table 2. WCA and WSA of flat and texturized surfaces, varying air exposure time, creep time, and template materials. [1]: 50 μL droplets.

Surface	Material	Air Exposure Prior to Polymerization (min)	Template Material	WCA (°)	WSA [1] (°)
Flat	Acrylate	0	Air	61.0 ± 1.3	–
	Acrylate	0	N_2	61.9 ± 2.3	–
	Acrylate	120	Air	71.8 ± 3.7	–
	Acrylate + 5 wt % PESiUA2	0	Air	104.5 ± 3.6	–
	Acrylate + 5 wt % PESiUA2	120	Air	101.5 ± 4.0	–
	PDMS	0	Air	112.7 ± 4.1	–
	Acrylate + 5 wt % PESiUA2	0	PDMS	91.4 ± 1.3	–
	Acrylate + 5 wt % PESiUA2	120	PDMS	90.7 ± 1.9	–
Nasturtium	Fresh leaf	–	–	143.9 ± 1.6	–
	Acrylate	0	PDMS	100.7 ± 3.6	–
	Acrylate + 5 wt % PESiUA2	0	PDMS	102.0 ± 2.1	–
	Acrylate + 5 wt % PESiUA2	120	PDMS	103.3 ± 3.2	–
Lotus	Fresh leaf	–	–	140.2 ± 0.9	–
	Acrylate	0	PDMS	139.1 ± 1.8	–
	Acrylate + 5 wt % PESiUA2	0	PDMS	143.9 ± 3.1	45
	Acrylate + 5 wt % PESiUA2	120	PDMS	141.7 ± 3.3	38
	Acrylate + 5 wt % PESiUA2	0	PDMS (5 min creep)	151.6 ± 1.0	30

Texturized surfaces were produced in a second step. The leaves of two plants were selected, namely lotus (*Nelumbo nucifera*) and nasturtium (*Tropaeolum majus*). The WCA of the lotus and nasturtium leaves, reported in Table 2, were found to be approximately 20° lower than previously reported values of 164° [25] and 162° [26], respectively, due to seasonal factors. The surface of the fresh leaves was used as master and was replicated in the formulations using a UV-nanoimprint lithography process (UVNIL) and an intermediate negative polydimethylsiloxane mold (PDMS, SYLGARD™ 184, Dow, Midland, MI, USA) as detailed in [17]. This soft, vacuum-free and ambient molding technique preserves the delicate biological surfaces from damage and enables to accurately reproduce their nanometer scale features [27,28]. In short, square samples (2 cm × 2 cm) cut from the leaves were fixed onto Petri dishes. PDMS was mixed with hardener (10:1 ratio). The mixture was homogenized manually for 5–10 min and degassed under a reduced pressure of 50.8 kPa for 5 min and 84.7 kPa for 10 min. It was poured onto the samples and subsequently cured at room temperature for 48 h. A UVNIL tool equipped with independent control of UV exposure and pressure was used to print the composite surface with the PDMS mold. A 200 μm thick layer of the liquid formulations was applied on a glass slide using a doctor blade. The PDMS mold was attached to another glass slide and the PDMS surface was put in contact with the liquid formulation under a controlled pressure of 3 bars. Curing was then performed using the same source and same UV dose as for the flat films. The printed surfaces were finally carefully demolded.

2.3. Characterization Methods

The water contact angle (WCA) of the polymerized surfaces was measured using a contact angle meter (EasyDrop, Krüss GmbH, Hamburg, Germany) at room temperature, with deionized water and a droplet volume of 10 μL. Four WCA measurements were made on each sample and the values were averaged. The water sliding angle (WSA) of selected surfaces was measured using a tilting support equipped with a protractor. A droplet of water was placed on the surface of the sample in horizontal position and the support was slowly tilted until the drop started to move. Measurements were made at room temperature using deionized water. The volume of the droplets varied from 10 to 100 μL.

The kinetics of the photopolymerization process were analyzed in real time during irradiation, by real-time Fourier transform infrared spectroscopy (RT-FTIR, Thermo-Nicolet 5700 spectrometer, Thermo Fisher, Waltham, MA, USA) on 12 μm thick coatings. The results are reported in the Supplementary Materials (Figure S1). Three formulations were tested: acrylate control, acrylate with 0.5 wt % of PESiUA2 and with 5 wt % of PESiUA2. The coatings were applied on a silicon wafer using a wire-wound Meyer bar; the IR experiments were made immediately or after 120 min exposure to air prior of starting the irradiation. The IR instrument was settled with a resolution of 4 cm^{-1} and the acquisition rate was 1 Hz. A 200 W high pressure mercury-xenon lamp (LC8, Hamamatsu, Shizuoka, Japan) was used and the intensity of the UV light was fixed at 66 mW·cm^{-2}. The conversion was calculated recording the decrease of the area of the absorption band of the C=C double bonds with time t, using the methacrylate peak at 1636 cm^{-1} ($P_{C=C}(t)$) [29]. The peak area of the C=O double bond at 1726 cm^{-1} ($P_{C=O}$) was chosen as the reference. The degree of conversion α was calculated as

$$\alpha = 1 - \frac{P_{C=C}\,(t)/P_{C=O}\,(t)}{P_{C=C}\,(0)/P_{C=O}\,(0)} \tag{1}$$

The influence of short flashes of UV light on the shear viscosity of selected formulations was determined using oscillatory shear rheometry and a UV-coupling cell (AR2000, TA instruments, New Castle, DE, USA), with a plate–plate geometry of diameter 20 mm and a gap of 500 μm. Samples were tested before, and after being illuminated for short periods. A strain amplitude sweep was performed first at 1 Hz to determine the limit for the linear viscoelastic range. Frequency sweep tests were then performed at a strain amplitude within the linear range, and usually close to 1%. The UV source and light intensity were the same as used for UVNIL experiments (75 mW·cm^{-2}). The rheology results are reported in the Supplementary Information.

The topography of the polymer surfaces was observed using a scanning electron microscope (SEM, FEI XLF30-FEG, Philips, Amsterdam, The Netherlands). The SEM was operated in high resolution mode using an acceleration voltage of 5 kV. The working distance was fixed to 10 mm. The samples were coated with a thin carbon layer of approximately 12 nm in order to avoid charging effects.

Chemical surface composition was evaluated by means of a PHI 5000 Versa-Probe scanning X-ray photoelectron spectrometer (XPS, PHI Versaprobe 5000, Physical Electronics, Inc., Chanhassen, MN, USA), with a monochromatic Al Kα source at 1486.6 eV. A spot size of 100 μm was used in order to collect the photoelectron signal for both the high resolution (HR) and the survey spectra. The semi-quantitative atomic compositions and deconvolution procedures were obtained using Multipack 9.7 dedicated software. All core level peak energies were referenced to C 1*s* peak at

284.5 eV (C–C) and the background contribution has been subtracted by means of a Shirley function. Depth profile has been performed using the Ar$^+$ source with a 2 kV ions accelerating voltage, alternate mode with sputter cycle of 30 s each.

3. Results and Discussion

3.1. Segregation of Siloxane Comonomers towards Polymer–Air Interface

The influence of the addition of the siloxane comonomers and their concentration on the wettability of the cured polymer surfaces at the air side was investigated by WCA measurements, in view of selecting the comonomer leading to the most hydrophobic surface, and the results are depicted in Figure 1. When photopolymerized onto the glass substrate, the WCA of the pure acrylate was 61°.

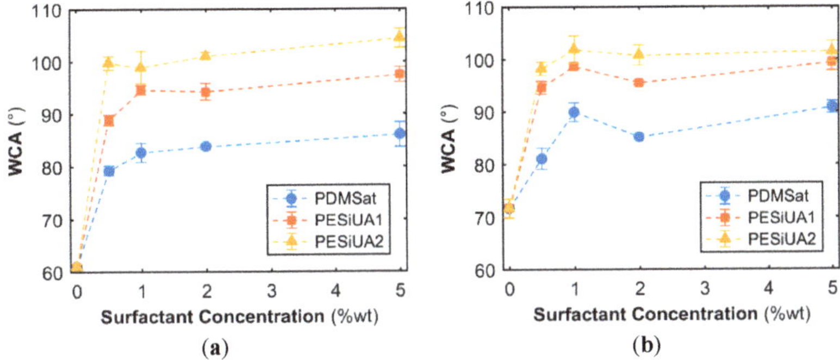

Figure 1. WCA vs. concentration of siloxane comonomer of flat surfaces polymerized without air exposure (**a**) and after 120 min of air exposure (**b**).

The addition of the three comonomers led to an increase of the WCA up to a saturation level around 1 wt %, beyond which the concentration increase had a marginal influence. Formulations with PESiUAs were beyond the hydrophobic limit independently of the concentration with WCA values comparable to the value for the PESiUAs homopolymers (104°). They were lower than the value of 113° for the PDMS SYLGARD™ 184 (Table 2), that is close to the maximum WCA of 115.2° for a flat surface [30]. In contrast PDMSat formulations only reached the hydrophobic limit after an exposure time of 120 min, although the WCA on PDMSat homopolymers were found to be equal to 97°.

Comparing the three siloxane comonomers, it is evident that the parameter influencing the surface activity of the comonomers was the length of the siloxane chain, i.e., the number of siloxane units. PDMSat contains only three units, which were not sufficient to impart a hydrophobic character. The dependence of wettability on the concentration and on the length of the apolar moiety of surface active comonomer has been already reported for fluorinated systems [17,31,32].

As detailed in Supplementary Materials (Figure S2), the delay time between coating application and irradiation processes was effective for the surface modification: comonomers diffused to the polymer–air interface, with a diffusion time depending on the siloxane structure. The higher the molecular weight the faster the diffusion, due to the reduced affinity with the bulk [32]: the PESiUA2 diffusion was immediate after coating, while PESiUA1 and PDMSat required longer times to reach the highest WCA values. In summary, PESiUA2 was the best choice to obtain the highest hydrophobicity of photocured films: it allowed obtaining the highest WCA without a delay time. However, the values were far away from the superhydrophobicity threshold (150°), therefore a change in surface morphology was required besides the modification of the chemical composition of the surface.

3.2. Influence of Texturization with PDMS Templates

The photopolymer showing the highest WCA for flat surfaces (acrylate + 5 wt % PESiUA2) was used to obtain texturized materials. The acrylate homopolymer was imprinted as reference, and Figure 2 shows the morphology of the synthetic replica of nasturtium and lotus leaves. The nasturtium replica was characterized by entangled island-like epithelial cells with dimensions around 100 μm [26]. The lotus replica exhibited a high density of micropapillae of diameter around 10 μm. A closer look revealed that the sub-micron epicuticular crystal structures were, however, not well replicated, in particular in the case of nasturtium. Such a difference in small-scale structures between the two types of plants was surprising since both natural surfaces are quite similar at this scale, with comparable tubular wax crystal morphologies. A first explanation is related to the possible erosion of the superficial waxes on old leaves [19], which would explain the rather low values of WCA on the nasturtium. In addition, the chemical composition of the lotus and nasturtium waxes differs totally [18], and one additional hypothesis would be that the nasturtium wax tubules partly dissolved in PDMS during the 48 h curing time [33]. Further work would be needed to clarify the observed differences.

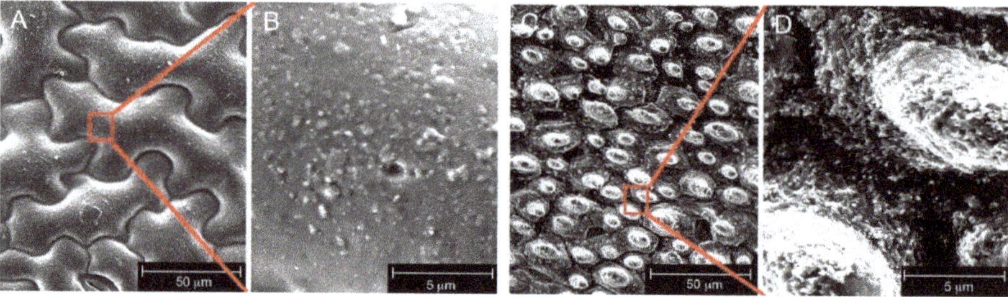

Figure 2. Electron micrographs of the synthetic replica made of photocured acrylate homopolymer: nasturtium (**A**) and higher magnification inset (**B**); lotus (**C**) and higher magnification inset (**D**).

Table 2 summarizes the WCA data for photopolymers having either flat surfaces or texturized surfaces. The WCA of the synthetic nasturtium surfaces with PESiUA2 comonomer was found to be slightly above 100°, as for the corresponding flat surface, and slightly higher than that of the synthetic nasturtium surface without comonomer. These values were 50° below the superhydrophobic threshold and some 40° below the WCA of the fresh nasturtium leaf. This result confirms that the nanosized epicuticular crystal features of the plant, which were not well replicated, are essential to achieve superhydrophobic properties. In contrast, the WCA of the synthetic lotus surfaces with PESiUA2 comonomer was found to be in the range 142°–144°, slightly higher than that of the fresh leaf and acrylate lotus surface, and very close to the superhydrophobic limit. In fact, the lotus replica was rather accurate (Figure 2C). This result further indicates the key role of the texture to promote a superhydrophobic state. Nevertheless, the WSA of the synthetic lotus surfaces also reported in Table 2 was rather large and much higher than the WSA of 3° of the plant, so that these synthetic replicas, although accurate were not superhydrophobic. Additional WSA data are reported in the Supplementary Materials (Table S1).

Surprisingly, the WCA of the flat surfaces with PESiUA2 comonomer polymerized in contact with a flat PDMS mold was 91°, i.e., below the values obtained without using the mold and irrespective of air exposure time prior to curing. This may imply that the contact with the PDMS template during the UV printing process led to a reverse migration of the comonomer as was reported for the case of fluorinated moieties [4]. All these measurements reveal the competing influence of PDMS and presence of siloxane monomer on the surface segregation of the latter.

Figure 3 shows XPS scans of the flat surfaces of acrylate and acrylate with PESiUA2 polymerized with and without 120 min of exposure to air prior to polymerization, and the lotus texturized surface of acrylate with PESiUA2 after 120 min of exposure to air. Apart from carbon and oxygen, present in all the surfaces, Si 2*p* and N 1*s* signals were carefully analyzed as fingerprints of the PESiUA2. The relative atomic concentration of these two elements is reported in Table 3. The aim was to first to confirm the migration of the PESiUA2 towards the polymer–air interface, and second to check the presence of this molecule at the surface of the texturized surface. The Si 2*p* and N 1*s* signals were indeed absent in the plain acrylate sample. The migration process was confirmed, with an increase of the superficial concentration of these two elements with time prior to polymerization, which is consistent with the increase of WCA with delay time shown in Figure 1. The concentration of Si at the surface of the lotus-texturized sample was found to be higher than that on the flat samples. Possible explanations are the sensitivity of the XPS measurement to surface roughness, and differences in superficial concentration of PESiUA due to different curing kinetics [34], hence different migration processes between the flat and the texturized materials.

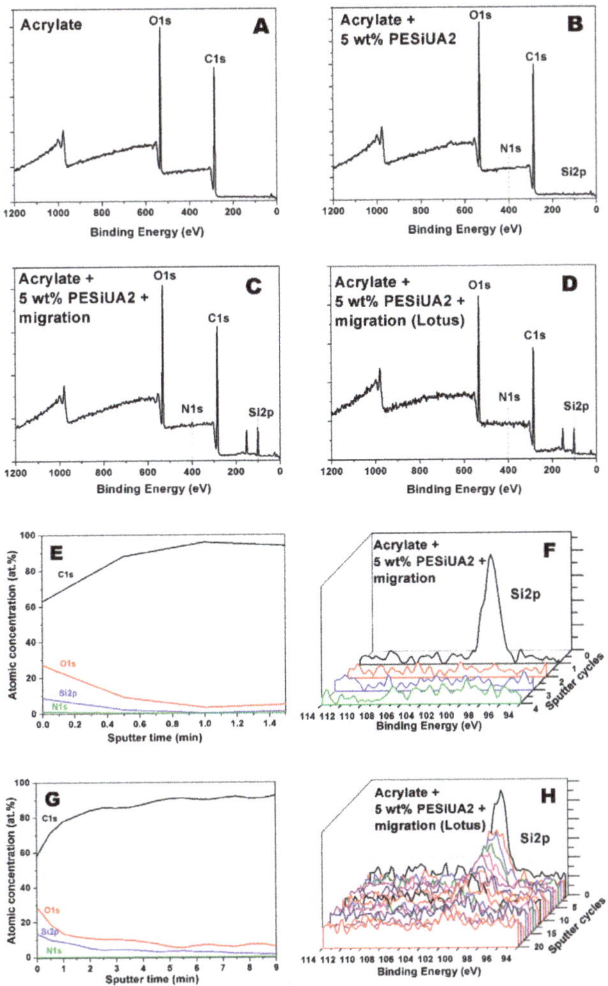

Figure 3. (**A–D**) XPS Survey scans for the four samples analyzed (see Table 3 for details). C 1*s*, O 1*s*, N 1*s*, and Si 2*p* peaks have been highlighted. Depth profiles of C 1*s*, O 1*s*, N 1*s*, and Si 2*p* of (**E**) Sample C and (**G**) Sample D. Si 2*p* high resolution curves vs. sputter cycles for (**F**) Sample C and (**H**) Sample D.

Table 3. Relative atomic concentration of Si and N at the surface of acrylate with and without PESiUA2, for flat and texturized surfaces.

Sample	Material	Air Exposure Prior to Polymerization (min)	Template Material	Si 2p (at.%)	N 1s (at.%)
A	Acrylate	0	air	0	0
B	Acrylate + 5 wt % PESiUA2	0	air	1.0	2.0
C	Acrylate + 5 wt % PESiUA2	120	air	8.5	1.3
D	Acrylate + 5 wt % PESiUA2	120	PDMS (Lotus)	12.7	0.9

The in-depth distribution of Si in the flat and texturized sample of acrylate with PESiUA2 polymerized after 120 min of exposure to air was further analyzed by depth-profiling using an Ar$^+$ source and is also shown in Figure 3. Assuming an etching rate of 5.5 nm/min for acrylate as previously determined for a BEMA–PEGMA siloxane enriched copolymer [35], we can see that there is an exponential decay of the Si 2p signal in the first 5 nm for the flat sample (Figure 3E), and in the first 10 nm for the texturized sample, beyond which the signal smoothly decreases till 50 nm in depth (Figure 3G). In Figure 3F,H, we have reported the Si 2p high resolution curves acquired during depth profiles for the two samples, to better highlight the signal decrease. Again, the differences in depth profiles may result from the sensitivity of the XPS measurement to surface roughness.

The XPS analyses imply that the siloxane comonomers did not fully migrate back into the bulk upon contacting the PDMS surface, in contrast with fluorinated comonomers [4]. To further check this, and totally prevent the trend for reverse migration, flashes of UV light were applied to the free surface of coatings after migration to chemically immobilize the surfactant before UVNIL. Acrylate coatings with 5 wt % of PESiUA2 exposed to air for 120 min were flashed for periods of 0.2 s up to 2 s before UVNIL with the negative PDMS lotus template. The WCA of the flashed, texturized surfaces turned out to be lower than that of surfaces produced without UV flash. As shown in the Supplementary Materials (Figure S3), this reduction of WCA was due to the large increase of the viscosity of the liquid formulation, by more than four orders of magnitude at low shear rates, and emergence of a yield stress behavior for flashes as short as 0.2 s. This considerable thickening compromised the fidelity of the low-pressure replication process. An alternative strategy was therefore tested, based on viscoelastic creep flow.

3.3. Influence of Creep

The low-pressure UVNIL process is fundamentally based on the viscoelastic creep flow of the resin into the sub-micron topography of the PDMS template, so that the replication is pressure and time-dependent. In order to allow the resin to better fill the mold cavities, a pressure of 3 bars was applied to the resin in contact with the PDMS mold for periods up to 10 min prior to UV curing. Figure 4 shows the influence of creep on the WCA of synthetic lotus surfaces, based on the acrylate with and without 5 wt % of PESiUA2 (SEM images of these surfaces are shown in Figure S4). In both cases, the WCA increased with creep to a maximum value of 152° (shown in Figure S5) with PESiUA2 after 5 min and then decreased. Nevertheless, the WSA reported in Table 2 was found to remain rather high and equal to 30°.

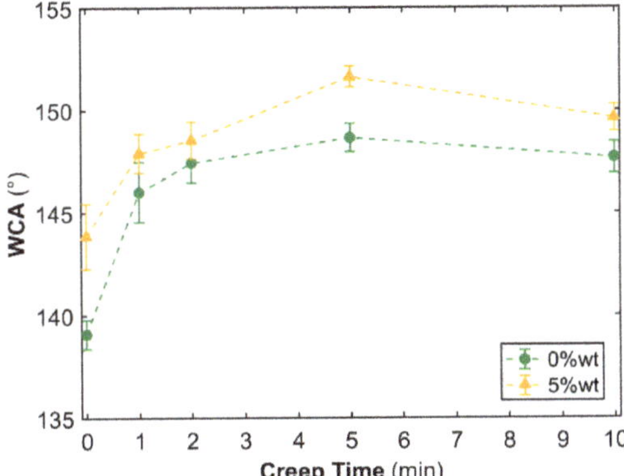

Figure 4. WCA of synthetic lotus surfaces based on acrylate and acrylate + 5 wt % PESiUA2 vs. creep time under 3 bars prior to UVNIL.

3.4. Self-Cleaning

Although the WSA was equal to 30° the synthetic acrylate + PESiUA2 lotus surface possessed effective self-cleaning properties. This was demonstrated by letting 10 μL water droplets fall from a height of 1 cm on the surface partly covered by ground pepper, on a glass support tilted at a 10° angle. A sequence of three images taken from a video are shown in Figure 5, where the capture of pepper grains by one bouncing droplet is evident (the video is available in the Supplementary Materials). Notice that, due to its hydrophobicity, ground pepper was reported to be more challenging to remove than other hydrophilic particles such as MnO and SiC [8]. The drop velocity upon impact was close to 0.4 m/s, which was high enough for the drop to bounce on the surface as observed (Figure 5b and Video S1 in the Supplementary Materials) [36]. The corresponding kinetic energy of the drop was close to 1 μJ, which again was high enough to overcome the effective interfacial energy E close to 0.07 μJ. The latter was roughly estimated as

$$E \sim \gamma_{sl} f \pi R^2 \tag{2}$$

where γ_{sl} is the water-polymer interfacial tension, found to be close to 40 mJ/m^2 from the measured WCA data, $f = (\cos\theta + 1)/(r\cos\theta_0 + 1)$ is the wet area fraction calculated and found to be equal to 13% using the Cassie–Baxter model [37] (θ and θ_0 represent the WCA of the texturized and flat surfaces, respectively, and r is the roughness factor, equal to 3.2 for lotus), and R is the radius of the droplet at the impact point, evaluated as 2 mm from the video images. Important to point out is that no such self-cleaning behavior was observed in the case of flat surfaces.

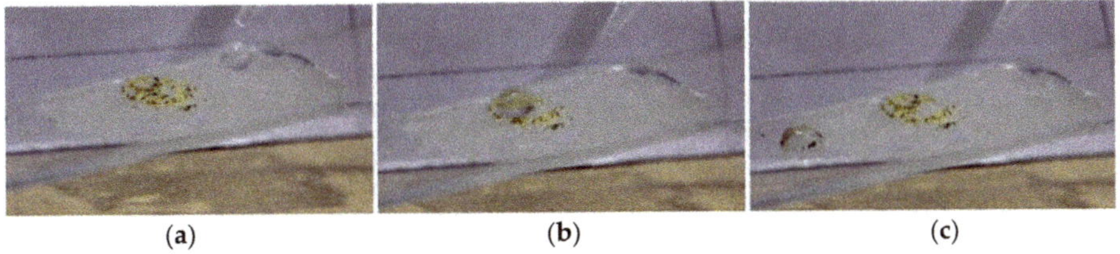

(a) (b) (c)

Figure 5. Sequence of photographs showing the behavior of a water droplet falling (**a**), bouncing (**b**), and rolling (**c**) on a synthetic acrylate lotus surface with 5 wt % of PESiUA2, printed on a glass support with 1 min of creep under 3 bars. The synthetic lotus surface was contaminated by pepper grains and support was tilted by 10°. Image (**b**) shows the bouncing droplet with trapped pepper grains, leaving a clean impact trace. Image (**c**) shows the water drop with trapped pepper grains, sticking on the smooth glass support surface after leaving the printed lotus surface.

The behavior of the synthetic lotus surface based on the acrylate with 5 wt % of PESiUA2 to other liquids than water is shown in Figure 6. The difference of contact angle is evident, confirming the hydrophobic character of the surface.

Figure 6. Coffee, olive oil, and water droplets from left to right on a synthetic lotus surface based on acrylate with 5 wt % PESiUA2 and 120 min air exposure prior to UVNIL.

4. Conclusions

Bio-inspired fluorine free synthetic replica of lotus and nasturtium surfaces were developed in view of obtaining self-cleaning properties, based on an energy efficient UV printing process. A UV-curable acrylate oligomer and three acrylated siloxane comonomers with different molecular weights were used. The process combined the self-assembly of the comonomers at the acrylate/air interface and a UVNIL replication step with a PDMS negative replica of the plant surfaces. The morphology and water contact angle of flat and texturized surfaces were systematically analyzed, leading to the following conclusions.

The migration of the comonomers towards the polymer–air interface led to an increase of the hydrophobicity of the surfaces, from 61° for the plain acrylate to 105° for the acrylate with 5 wt % of PESiUA2 after 30 min of exposure to air prior to photopolymerization. Comonomers with a higher molecular weight migrated faster.

The replication fidelity was excellent for the lotus leaf surface whose hierarchical texture comprising micropapillae and sub-micron crystalloids was well reproduced in the acrylate/comonomer material. It was less accurate for the nasturtium surface, with a lack of replication of the sub-micron features. The texturization of the surfaces had a large influence on hydrophobicity. The WCA of synthetic replica of lotus and nasturtium with 5 wt % of PESiUA2 was equal to 144° and 102°, respectively. The highest WCA, equal to 152° was measured on a synthetic lotus surface with 5 wt % PESiUA2 and 5 min of creep flow under a pressure of 3 bars prior to polymerization. The WSA of this surface was found to be equal to 30° due to weak adhesion of water. The surface was nevertheless self-cleaning with water droplets when contaminated with hydrophobic pepper particles, provided that the droplets had some kinetic energy.

Supplementary Materials: The following are available online at http://www.mdpi.com/2079-6412/8/12/436/s1, Figure S1: Photoconversion vs. time for HBP and HBP-PESiUA2 (0.5 and 5 wt %) without and with 120 min of air exposure prior to measurements; Figure S2: WCA vs. air exposure time of flat surfaces of acrylate and acrylate-siloxane formulations: (a) PDMSAT; (b) PESiUA1; and (c) PESiUA2; Figure S3: Viscosity of acrylate (a), acrylate + 0.5 wt % PESiUA2 (b), acrylate + 5 wt % PESiUA2 (c) as function of angular frequency, for increasing UV flash duration (0.2–2 s) as indicated; Figure S4: Scanning electron micrographs of the synthetic lotus surfaces, based on the acrylate with 5 wt % PESiUA2, without (a) and with (b) a creep time of 5 min under a pressure of 3 bars prior to photopolymerization; Figure S5: WCA of 151.9° measured on a synthetic lotus replica surface with 5 wt % PESiUA2 and 5 min of creep; Table S1: WSA of synthetic lotus surfaces; Video S1: MVI_8122.mv4.

Author Contributions: Conceptualization, S.D.V. and Y.L.; Data Curation, L.W., L.M., and M.C.; Formal Analysis, L.W., S.D.V., F.K., L.M., M.C., A.V., R.B., and Y.L.; Funding Acquisition, Y.L.; Investigation, L.W., S.D.V., F.K., A.V., R.B., and Y.L.; Methodology, R.B.; Supervision, S.D.V., R.B., and Y.L.; Writing—Original Draft Preparation, L.W. and M.C.; Writing—Review & Editing, S.D.V., F.K., L.M., A.V., R.B., and Y.L.

Funding: This research was partly funded by EPFL's Integrated Food and Nutrition Center.

Acknowledgments: The authors acknowledge the Botanical Garden of Lausanne for the supply of fresh plants.

Conflicts of Interest: The authors declare no conflict of interest.

References

1. Shirtcliffe, N.J.; McHale, G.; Newton, M.I.; Chabrol, G.; Perry, C.C. Dual-scale roughness produces unusually water-repellent surfaces. *Adv. Mater.* **2004**, *16*, 1929–1932. [CrossRef]
2. Bhushan, B.; Jung, Y.C. Natural and biomimetic artificial surfaces for superhydrophobicity, self-cleaning, low adhesion, and drag reduction. *Prog. Mater. Sci.* **2011**, *56*, 1–108. [CrossRef]
3. Wong, T.-S.; Kang, S.H.; Tang, S.K.Y.; Smythe, E.J.; Hatton, B.D.; Grinthal, A.; Aizenberg, J. Bioinspired self-repairing slippery surfaces with pressure-stable omniphobicity. *Nature* **2011**, *477*, 443–447. [CrossRef] [PubMed]
4. Roach, P.; Shirtcliffe, N.J.; Newton, M.I. Progress in superhydrophobic surface development. *Soft Matter* **2008**, *4*, 224–240. [CrossRef]
5. Ming, W.; Wu, D.; van Benthem, R.; de With, G. Superhydrophobic films from raspberry-like particles. *Nano Lett.* **2005**, *5*, 2298–2301. [CrossRef] [PubMed]
6. Budunoglu, H.; Yildirim, A.; Guler, M.O.; Bayindir, M. Highly transparent, flexible, and thermally stable superhydrophobic ORMOSIL aerogel thin films. *ACS Appl. Mater. Interfaces* **2011**, *3*, 539–545. [CrossRef] [PubMed]
7. Mates, J.E.; Ibrahim, R.; Vera, A.; Guggenheim, S.; Qin, J.; Calewarts, D.; Waldroup, D.E.; Megaridis, C.M. Environmentally-safe and transparent superhydrophobic coatings. *Green Chem.* **2016**, *18*, 2185–2192. [CrossRef]
8. Schutzius, T.M.; Bayer, I.S.; Qin, J.; Waldroup, D.; Megaridis, C.M. Water-Based, Nonfluorinated dispersions for environmentally benign, large-area, superhydrophobic coatings. *ACS Appl. Mater. Interfaces* **2013**, *5*, 13419–13425. [CrossRef] [PubMed]
9. Olin, P.; Hyll, C.; Ovaskainen, L.; Ruda, M.; Schmidt, O.; Turner, C.; Wågberg, L. Development of a semicontinuous spray process for the production of superhydrophobic coatings from supercritical carbon dioxide solutions. *Ind. Eng. Chem. Res.* **2015**, *54*, 1059–1067. [CrossRef]
10. Smith, J.D.; Dhiman, R.; Anand, S.; Reza-Garduno, E.; Cohen, R.E.; McKinley, G.H.; Varanasi, K.K. Droplet mobility on lubricant-impregnated surfaces. *Soft Matter* **2013**, *9*, 1772–1780. [CrossRef]
11. Schlaich, C.; Yu, L.; Cuellar Camacho, L.; Wei, Q.; Haag, R. Fluorine-free superwetting systems: Construction of environmentally friendly superhydrophilic, superhydrophobic, and slippery surfaces on various substrates. *Polym. Chem.* **2016**, *7*, 7446–7454. [CrossRef]
12. Torstensson, M.; Ranby, B.; Hult, A. Monomeric surfactants for surface modification of polymers. *Macromolecules* **1990**, *23*, 126–132. [CrossRef]
13. Bongiovanni, R.; Malucelli, G.; Priola, A. Modification of surface properties of UV-cured films in the presence of long chain acrylic monomers. *J. Colloid Interface Sci.* **1995**, *171*, 283–287. [CrossRef]
14. Van der Grinten, M.G.D.; Clough, A.S.; Shearmur, T.E.; Bongiovanni, R.; Priola, A. Surface segregation of fluorine-ended monomers. *J. Colloid Interface Sci.* **1996**, *182*, 511–515. [CrossRef]
15. Bongiovanni, R.; Sangermano, M.; Medici, A.; Tonelli, C.; Rizza, G. Nanostructured hybrid networks based on highly fluorinated acrylates. *J. Sol-Gel Sci. Technol.* **2009**, *52*, 291–298. [CrossRef]
16. Sangermano, M.; Bongiovanni, R.; Longhin, M.; Rizza, G.; Kausch, C.M.; Kim, Y.; Thomas, R.R. Hybrid organic/inorganic UV-cured acrylic films with hydrophobic surface properties. *Macromol. Mater. Eng.* **2009**, *294*, 525–531. [CrossRef]
17. González Lazo, A.M.; Katrantzis, I.; Dalle Vacche, S.; Karasu, F.; Leterrier, Y. A Facile in situ and UV printing process for bioinspired self-cleaning surfaces. *Materials* **2016**, *9*, 738. [CrossRef] [PubMed]
18. Koch, K.; Dommisse, A.; Barthlott, W. Chemistry and crystal growth of plant wax tubules of lotus (*nelumbo nucifera*) and nasturtium (*tropaeolum majus*) leaves on technical substrates. *Cryst. Growth Des.* **2006**, *6*, 2571–2578. [CrossRef]
19. Neinhuis, C.; Barthlott, W. Characterization and distribution of water-repellent, self-cleaning plant surfaces. *Ann. Bot.* **1997**, *79*, 667–677. [CrossRef]

20. Schmidt, L.E.; Schmah, D.; Leterrier, Y.; Manson, J.A.E. Time-intensity transformation and internal stress in UV-curable hyperbranched acrylates. *Rheol. Acta* **2007**, *46*, 693–701. [CrossRef]

21. Schmidt, L.E.; Yi, S.; Jin, Y.-H.; Leterrier, Y.; Cho, Y.H.; Månson, J.A.E. Acrylated hyperbranched polymer photoresist for ultra-thick and low-stress high aspect ratio micropatterns. *J. Micromech. Microeng.* **2008**, *18*, 045022. [CrossRef]

22. Geiser, V.; Jin, Y.H.; Leterrier, Y.; Manson, J.A.E. Nanoimprint lithography with UV-curable hyperbranched polymer nanocomposites. *Macromol. Symp.* **2010**, *296*, 144–153. [CrossRef]

23. Geiser, V.; Leterrier, Y.; Månson, J.-A.E. Low-stress hyperbranched polymer/silica nanostructures produced by uv-curing, sol-gel processing and nanoimprint lithography. *Macromol. Mater. Eng.* **2012**, *297*, 155–166. [CrossRef]

24. Cheng, J.; Li, M.; Cao, Y.; Gao, Y.; Liu, J.; Sun, F. Synthesis and properties of photopolymerizable bifunctional polyether-modified polysiloxane polyurethane acrylate prepolymer. *J. Adhes. Sci. Technol.* **2016**, *30*, 2–12. [CrossRef]

25. Bhushan, B. Biomimetics: Lessons from nature—An overview. *Philos. Trans. Royal Soc. A Math. Phys. Eng. Sci.* **2009**, *367*, 1445–1486. [CrossRef] [PubMed]

26. Sharma, C.S.; Abhishek, K.; Katepalli, H.; Sharma, A. Biomimicked superhydrophobic polymeric and carbon surfaces. *Ind. Eng. Chem. Res.* **2011**, *50*, 13012–13020. [CrossRef]

27. Sun, M.; Luo, C.; Xu, L.; Ji, H.; Ouyang, Q.; Yu, D.; Chen, Y. Artificial lotus leaf by nanocasting. *Langmuir* **2005**, *21*, 8978–8981. [CrossRef] [PubMed]

28. Lee, S.-M.; Kwon, T.H. Effects of intrinsic hydrophobicity on wettability of polymer replicas of a superhydrophobic lotus leaf. *J. Micromech. Microeng.* **2007**, *17*, 687. [CrossRef]

29. Vitale, A.; Quaglio, M.; Cocuzza, M.; Pirri, C.F.; Bongiovanni, R. Photopolymerization of a perfluoropolyether oligomer and photolithographic processes for the fabrication of microfluidic devices. *Eur. Polym. J.* **2012**, *48*, 1118–1126. [CrossRef]

30. Nakajima, A. Design of hydrophobic surfaces for liquid droplet control. *NPG Asia Mater.* **2011**, *3*, 49–56. [CrossRef]

31. Mikhaylova, Y.; Adam, G.; Häussler, L.; Eichhorn, K.-J.; Voit, B. Temperature-dependent FTIR spectroscopic and thermoanalytic studies of hydrogen bonding of hydroxyl (phenolic group) terminated hyperbranched aromatic polyesters. *J. Mol. Struct.* **2006**, *788*, 80–88. [CrossRef]

32. Bongiovanni, R.; Di Meo, A.; Pollicino, A.; Priola, A.; Tonelli, C. New perfluoropolyether urethane methacrylates as surface modifiers: Effect of molecular weight and end group structure. *React. Funct. Polym.* **2008**, *68*, 189–200. [CrossRef]

33. Lee, J.N.; Park, C.; Whitesides, G.M. Solvent Compatibility of poly(dimethylsiloxane)-based microfluidic devices. *Anal. Chem.* **2003**, *75*, 6544–6554. [CrossRef] [PubMed]

34. Vitale, A.; Touzeau, S.; Sun, F.; Bongiovanni, R. Compositional gradients in siloxane copolymers by photocontrolled surface segregation. *Macromolecules* **2018**, *51*, 4023–4031. [CrossRef]

35. Sacco, A.; Bella, F.; De La Pierre, S.; Castellino, M.; Bianco, S.; Bongiovanni, R.; Pirri Candido, F. Electrodes/electrolyte interfaces in the presence of a surface-modified photopolymer electrolyte: Application in dye-sensitized solar cells. *ChemPhysChem* **2015**, *16*, 960–969. [CrossRef] [PubMed]

36. Richard, D.; Clanet, C.; Quéré, D. Contact time of a bouncing drop. *Nature* **2002**, *417*, 811. [CrossRef] [PubMed]

37. Cassie, A.B.D.; Baxter, S. Wettability of porous surfaces. *Trans. Faraday Soc.* **1944**, *40*, 546–551. [CrossRef]

MDPI

St. Alban-Anlage 66

4052 Basel

Switzerland

Tel. +41 61 683 77 34

Fax +41 61 302 89 18

www.mdpi.com

Coatings Editorial Office

E-mail: coatings@mdpi.com

www.mdpi.com/journal/coatings